Earth Ways

Earth Ways

Framing Geographical Meanings

Edited by Gary Backhaus and John Murungi

LEXINGTON BOOKS
Lanham • Boulder • New York • Toronto • Oxford

LEXINGTON BOOKS

Published in the United States of America
by Lexington Books
An imprint of The Rowman & Littlefield Publishing Group, Inc.
4501 Forbes Boulevard, Suite 200, Lanham, Maryland 20706

PO Box 317
Oxford
OX2 9RU, UK

British Library Cataloguing in Publication Information Available

Library of Congress Cataloging-in-Publication Data

Earth ways : framing geographical meanings / edited by Gary Backhaus and John Murungi.
p. cm.
Includes bibliographical references and index.
ISBN 0-7391-0764-X (alk. paper) ISBN 978-0-7391-7064-5
1. Geography—Philosophy. 2. Geographical perception. I. Backhaus, Gary, 1953– II. Murungi, John, 1943–
G70.E27 2004
910'.01—dc22 2003021920

Printed in the United States of America

™ The paper used in this publication meets the minimum requirements of American National Standard for Information Sciences—Permanence of Paper for Printed Library Materials, ANSI/NISO Z39.48–1992.

Contents

III. Framing Case Studies of Specific Time-Places

Preface

In the *Republic*, Plato declared that justice consists of everyone minding his or her own business and not being a busybody. But wasn't Socrates' divine calling that of a busybody, asking people questions and prying into their business? Academia seems to have taken Plato's advice with a vengeance. Academicians of different disciplines stick to their own business and only to that business. And it seems that the academy has also adopted Plato's method of division, for, as disciplines tend toward increasing specialization, people in the same discipline become partitioned with each sub-discipline training its members to mind their own business. If this trend keeps up, then every single individual will be doing his or her own work and minding his or her own business. Much of academic research appears to be conducted in this solipsistic manner.

The conveners of the Annual International Conferences at Towson University have made it a primary goal to engage in the activities of busybodies, the ways of the unjust, in Plato's estimation. Our conference participants are not content with minding the business of their own separate disciplines. Rather, we seek to de-center ourselves by forming new relations and new syntheses rather than further divisions. We intend to subvert the individualism of academics, the lone thinkers, and to reestablish the Socratic symposium, which thrives through collective conversation—a veritable sharing and engagement.

But perhaps the divisions of academia are of the erroneous type like Plato's example of dividing cobblers on the basis of those who have retained their hair and those who are bald. Is it the case that the only academicians who can think are philosophers? Is it the case that the only academicians who have an aptitude for the spatial configurations on the earth's surface are geographers? These capacities, functions, and concerns can and are to be shared, and each discipline can engage in thinking about the spatial manifestations of phenomena. The various disciplines and subdisciplines bring their own fresh perspectives to the concerns of philosophy and geography. So, it has been the business of the co-editors to invite everyone to join us in the landscape of busybodydom!

Earth Ways reflects the title of the Fourth International Annual Philosophy and Geography Conference, which was held in the year 2002 at Towson University, Towson, Maryland, and from which many of the following chapters have been collected. *Earth Ways* carries the sense of the manner, style, or, more technically, organizing processes as functions of both the activity of knowing and the landscape. In preparing their chapters, the authors were asked to thematically address the organizing component, or *framings*, that manifested through bridging from their own disciplines to philosophy and geography. We have named these meaning-contexts, "frames." These frames initially were unthematized aspects of the authors' concerns. The co-editors requested that these frames be made thematic in the interest of creating a coherent interdisciplinary volume. Thus the authors were asked to reflect specifically upon their transdisciplinary turn, which emphasized their bridge to geographical and philosophical concerns. What the co-editors realized was that by making this reflective effort to cross disciplines, a reorganization of the contents emerged. So, in selecting essays for this volume the co-editors dialogued with the essayists to help them make a shift of focus in order for the geographical organization, the spatial moment, of the contents of their research to become their central concern. Truly creative interdisciplinary ideas have emerged from this project. The volume exhibits a flexibility that invites readers to frame what they read in ways that bring to fruition what is pregnant in the text. Although it is our objective to organize a coherent text, it is anticipated that the reader will join in this effort. Even what may appear to be inconsistent or open to challenge has an important chaotic role for the generation of new coherences in the subtexts of each of the chapters. These destructurings perhaps will bring about new framings in the minds of our readers, which may then motivate their own earth way projects!

We would like to thank the many people at Towson University, especially Dr. Beverly C. Leetch, the Dean of the College of Liberal Arts, for continuing to support our annual international philosophy and geography conferences. We would like to thank everyone who has participated in our conferences, especially those who have been attending regularly and those who have contributed to our volumes. We would like to thank Jason Hallman, Acquisitions Editor for Lexington Books, for his commitment to this form of study and his continued support of our projects. We also appreciate the printing and computer expertise of Jeep Watson, who helps with our poster and book-figure designs, and also of Joanne Anderson, who has created and continues to organize our webpage. Please visit our webpage: http://wwwnew.towson.edu/spg and please see our other volumes: *Transformations of Urban and Suburban Landscapes: Perspectives from Philosophy, Geography, and Architecture* and *Tensional Landscapes: The Dynamics of Boundaries and Placements*, also published by Lexington Books. We look forward to your participation in our annual conferences!

Introduction

Earth Ways: The Primordial Relation Between the Ways of Knowing and the Ways of Earthly Phenomena

Gary Backhaus

Earth Ways—a pleasing poetic ring to the phrase that titles this volume of studies. Once we have tasted of its sensuous savor, we soon ask after its significance. "Way" is meant to indicate a modality of Being, a characteristic of an earthly being's concrete and contextualized existence. The many earthly beings, with their multifarious modalities of Being primordially belong to the Being-of-the-Earth, and make their way(s) only through the sustaining power of the Earth that supports them. The ways of an entity always imply this primordial relation to the earth, which are carved out in relation to a nexus of ways that sustain the many other earthly beings. All ways of Being are contextualized and are ultimately bound to the Earth's horizon as the primordial context or Way.

Being-of-the-Earth, as the primordial source for all ways of beings, or the Being of beings, cannot be framed. Framing issues from being-situated, already contextualized in the Earth's horizon. Being-of-the-Earth is the ground from which structures and processes issue forth, which are the frames of beings that can be framed through modalities of knowing. Contextualized in those structures and processes, humankind seeks to understand those earthly modalities, to articulate them in language, which gives rise to various worldviews. Worldviews, as coherent meaning-contexts, establish relations to the Earth, a style of Being-of-the-Earth. A worldview is like a posture of the lived-body; it is not the lived-body itself, but a way of engagement in the world. We are always situated within the Being-of-the-Earth and thus our worldviews are Earthbound postures.

A particular earthly being exhibits a multitude of ways, which are only made present to our awareness through the ways-by-which-we-engage that earthly being. We delimit our inquiry by recognizing that our interest in earthly ways includes and entails *ways of knowing.* We are not simply attending to earthly ways or to ways of knowing. Knowing modalities provide channels to apprehend certain features of earthly beings. Since, as we have already indicated, ways always involve a relational primordiality, we set our interests upon ways of knowing *in relation to* the ways of the earthly. The activity of knowing, which is itself an earthly way of a particular type of earthly being, always involves a relation to its content, something known. A being whose Being includes the modality of knowing as one of its capacities, exhibits the potential to apprehend the various modalities (ways) of the multifarious earthly beings. Knowing itself involves modalities, or multitudinous channels by which the various features of earthly beings can be brought to articulation. The ways of knowing are further delimited by our study: we are specifically engaged in the *framing of geographical meanings*, which alludes to the subtitle of this volume. A particular component of earthly ways, the geographical significance of an earthly being, is to be apprehended through a particular way of knowing, "framing."

Let us first illustrate what we mean by a delimitation put on "earth ways" by discussing the historical emergence of a particular way of knowing. The dawn of Western scientia is said to have occurred in the Ionian port city of Miletus, Asia Minor, in the sixth century before the Common Era. The usual procedure of study is to interpret the writing fragments of the Milesians functionally, as answers to questions, from which scholars determine the philosophical problems that the Milesians would have been pondering. Let us for our purpose pose an additional *way* and observe that the relationships of thought (knowing) to experience, thought to reality, experience to reality, and the triadic structure of thought-experience-reality are already of issue, even if there was yet not much thematic awareness of these epistemological problematics. These relationships in the modality of knowing are fundamental components of human existence, and so are always in relation—we are always aware of being aware of something, which is the fundamental condition—spontaneous reflection—for being a consciously knowing being. Because of the metaphysical nature of the Milesians' knowledge claims, these epistemological relations beckon us to pay attention to them. By what *way* did these thinkers arrive at their knowledge? These first philosophical inquiries into the nature of reality seem to involve abstractions from the experience of the earth's primary elements of water, air, fire, and earth. These objects of experience, together or individually, were nominated as the fundamental source(s) of all things through observing their specific qualities. In these earliest doctrines it is apparent that the natural landscape is observed and some element(s), due to properties/structures (ways/modalities) that are apprehended, become(s) thematic in reflection and subsequently interpreted as the derivative source for the many things that we experience. Some species of metaphysical speculation, as a *way of knowing*, reflect on an element's observable characteristics and propose that the *way of derivation* is the primordial process that accounts for the multifarious ways of the Many. For example, Anaximenes explains the processes of condensation and rarefaction by noting that breath blown out from compressed lips is cool, but if the mouth is opened wide, the breath is warm. This observation allows him to explain *the way* by which the many things are derived from air—the source of the myriad of beings. Yet, a rival school of thought promotes that the many things that we experience have emerged through *the way of separation* from a mixture of irreducible opposites. We can thus notice that emphasis is placed not only on different elements and their characteristics, but also on organizing processes, whether derivational or separational. And in combination of elements and processes, a relational nexus manifests as the important aspect in the *ways of primordial cosmology*—the relation of the Many (experienced entities) to the One (universe).

Observation of the natural world not only spawned speculation in the cosmogony of *the One*; the Milesians were interested in *the Many* for its own sake. The many entities of the earth were encountered through geographical explorations by which the Milesians observed differences in peoples and in the physical nature of the earth. "Hekataios of Miletus (ca. 520 B.C.) wrote the treatise, *Circuit of the Earth*, and sketched a map on which details of voyages, coastlines, cities, and peoples could eventually be plotted."[1] Both the Milesian metaphysical speculation of the One and their geographical exploration of the Many involved a keen *observation of forms—ways of beings*, which required collecting, discriminating, and classifying phenomena, *ways of knowing*. I shall later have more to say about *the way of formism* as a fundamental way of knowing.

Parmenides was the first thinker to radically partition thought from experience. Through the logical requirements that follow from his characterization of Being, the objects of the *way of reason* are said to be true and the objects of the *way of the senses* are taken as merely appearances. Reality can only be of thought and cannot be experienced. After Parmenides the problem of the One and the Many becomes the problem of how "to save" the way of earthly appearances. How can the logical demands of human reason be reconciled with the existence of the many beings that we experience? This question is perennial to philosophy. Both Parmenides and the Pythagoreans had influence on Plato's epistemological doctrine to remove reality from the earth as an otherworldly level transcending sense experience—the ideal realm of the forms. Commenting on the repercussions of Plato's intellectualism for the *way of designing built spaces*, architect Kimberly Dovey states, "Plato declared in the *Timaeus* that geometry should be the science of space. Subsequent developments through Euclid have built geometric space into a powerful model for understanding the world. This model is so powerful that a geometric representation of the world is widely perceived as the one true arrangement."[2]

In *World Hypotheses*, philosopher Steven Pepper describes four fundamental Western worldviews (ways of discerning, organizing, and producing knowledge) based on root metaphors: formism, mechanism, organicism, and contextualism.[3] What I have discussed so far about the Greeks concerns the worldview (the way of understanding reality) of formism. Geographer Anne Buttimer distinguishes two kinds of formism: immanent and transcendent. "Immanent formism implies a straightforward description, via symbolic representation, of phenomena. It identifies (1) characters, (2) particulars, and (3) levels of participation, or the tie between characters and particulars."[4] "Transcendent formism went beyond the level of description and representation. It sought *norms*, or models, yielding explanation of patterns."[5] These two senses of formism are present in the Greek tradition, and these two senses of formism are in tension, which subsequently becomes a technical problem for Greek cartographers.

> Mapmakers from the earliest times faced a twofold task: one was to represent phenomena symbolically as illustration and grounding for *historia* (à la Heroditus); the other challenge was to devise a mathematical based grid system to serve as a potentially universal framework for the study of the earth or the cosmos as a whole (à la Ptolemy). While the former maintained an inductive approach and aim at contextual sensitivity to local detail and contextual variety among the phenomena observed, the latter pointed toward abstract models of the earth and a deductive science of location and spatial organization."[6]

The epistemological bridge that would establish the relation between these two ways of knowledge is missing, and this lacuna has haunted western scientia ever since.

This brief illustration of Greek philosophy and geography and the discussion of its dominant way of knowing—formism—is sufficient to help make more sense of what is being meant by earth ways. "Earth ways" is not to be taken as an objective phenomenon describing some sort of entity that is separate from humans, which rests on the inconsistent idea that humanity or human observation is not also an earth way. "Earth ways" entails earthly beings whose ways of knowing involve apprehending the multitudinal ways of the plethora of earthly phenomena. The reflective apprehension of ways of earthly beings is not of pure immediacy; it constitutes a mediated sphere, which is called knowledge. Ways of knowledge apprehend ways of earthly beings by forming special relations with those ways. John Kirtland Wright invented a kindred notion to earth ways in his idea of geosophy, which could be described as the study of the collusion of geography and worldviews.[7] It is the geosophical worldviews or earth ways that constitute the "source of many paradigms, models, and analogues that have guided analytical frontiers of science and the humanities."[8] Apprehension of an earth way involves a particular engendering way (worldview) that delimits the activity of a knower for disclosure of that earth way. A particular way of an earthly being is summoned through a particular way of knowing. So "Earth ways" entails the unity of the experiencer and the experienced, and a reflective experiencing or way of knowledge channels what way that an earthly entity presents itself.

Ways of Knowing as Channeling Access to the Ways of Phenomena

Steven Pepper's four worldviews provide four ways of deriving and interpreting facts, "each ascribing to a distinct theory of truth and generating a set of categories through which its discourse on reality was to be *channeled*."[9] [emphasis added.] But only thinkers working at the most general levels consciously propound a worldview. Much of the time worldviews continue to be generated tacitly by methodological procedures that operationally set channels, which predetermine the kind of ways apprehended. This tendency of forgetting the foundation turns into operational hegemony: the character of a phenomenon is summoned through a particular way and this operational presentation is taken as *the* character of reality. Thus it is highly important and relevant to recognize the function of worldviews in establishing methodologies. Socrates turned his attention to the knower and advocated the idea that reality can only be grasped by properly training the soul. His methodology for attaining reality involves dialectical thinking and discourse. Aristotle advocated contemplation as the highest calling, for such a methodology most resembled the way of the Unmoved Mover. René Descartes ushers in the modern epoch by making the very thinking of the thing that thinks the foundation of scientia. For Descartes, pure rational thought following the strictest methodology of rules provides access to the quantitative mechanical reality behind the veil of qualitative experience.

We engage in this meta-theoretical attention to the ways of knowing to draw attention to the fact that a particular methodological way limits the way that a phenomenon can be apprehended. There is a further important reason to understand this point. In guiding the essayists included in this volume, we, the co-editors, have advocated a way of knowing, *framing*, such that a particular way of earthly phenomena is disclosed. We can now begin delineating this way.

Thomas Kuhn's term, "paradigm shift," can refer to an "earthshaking" discovery in scientia whereby one of the various worldviews gains a position of hegemony.[10] As the prime "catalyst" of the modern worldview, Descartes supplied the philosophical foundation for a mechanical universe (Pepper's mechanism), which was implied in the experimental work of Galileo. The Galilean way served as the exemplar for the methodologies upon which the modern sciences generated their knowledge-contents. The methodologies that developed across various regions of phenomenon sought the same goal: to discover the mechanical causes *behind* phenomena. Mechanical laws that govern phenomena exhibit operational or manipulative, predictive, and controlling values. The central *way of knowing* is to explain the qualitative phenomena of observable appearances through isolating causal chains that are to be measured by constructing a quantifiable index. Mechanical explanation connects that which is an observable appearance to that which is unobservable reality. The presupposition of this understanding can be called the double world theory. There is a mechanical world (reality) "behind" the qualitative world of experience (merely subjective appearances), which is accessed through reducing the world of experience to purely quantitative analyses. Explanation involves demonstrating how the qualitative world that we experience is caused to appear on the basis of this purely mechanical world of atoms in homogeneous space.

No culture in the history of humankind other than the modern Europeans ever thought to turn the world into a machine—the quantitative mechanical universe. This reductive *way of knowing* became the dominant worldview and the results are at once fantastic and horrific. The growth and success of the natural sciences advocated the further attempt to bring the knower into the phenomena that is to be scientifically studied—as the "tag-end" of causal chains. The attempt was to show that humans, the knowers, are nothing more than machines as well, and their Being is equally reducible to the mechanical way. The absurdity is then that nothing really has any value or any meaning, for everything in reality is nothing more than quantifiable material. The result of the hegemony of the mechanical worldview was to partition science from life and to diminish the meaning of the lifeworld to merely subjective and unimportant surface appearances, which is to posit scientific abstraction as true reality. Not until the twentieth century did philosophers reveal the *scientific way of quantitative abstraction* as a crisis in European culture and attempt to re-establish the primacy of lifeworld. The ongoing struggle is to relink scientia to the lifeworld as its foundation instead of following the way that serves the abstractions that supposedly explain the lifeworld and devalue it as the non-reality of merely subjective experiences. Having made these strong accusations against the double world theory of mechanism as a dominating component in its worldview, especially as tacitly presupposed in the developed operationalisms in its sciences, it must be noted that mechanics has a role to play as a way of human knowledge. And, with a philosophical understanding of the limits of its epistemological and ontological scope, mechanics can help improve the world without appropriating it.

We have discussed mechanics for the purpose of contrasting its ways (pardigms, models, and analogues) with the way of knowing that we call "framing," which belongs to the tradition of formism. We do not advocate formism as the worldview that should dominate or serve as foundation for others. Our goal, which grounds the discussions of the authors who have written chapters in this volume, will be to situate this way of knowing (framing) in the many ways of knowing and to elucidate its fundamental significance.

Framing Geographical Meanings

The term, "framing" was suggested to us from perhaps an unusual source—sociologist Erving Goffman's *Frame Analysis*.[11] By "frame" Goffman means a scheme of interpretation by which objects, events, and activities organize so that we make sense of them—interpret their meaning. The meanings of phenomena are dependent upon the framework from which they are perceived. We don't intend to use the word "frame" as a subjective category, or for that matter as an objective category. Epistemologically we understand framing as entailing the subject-object relation as a primordial unity. "Framing" entails a discernable, qualitatively noetic (knowing) activity, a type of organizing activity by which a frame of phenomena (structurization) is apprehended. Following Husserl we could say, an act of framing apprehends (intends/constitutes) a frame-object. Or better, an act of type A framing apprehends a type A frame-object, for we intend a particular kind of framing, the geographical. Not only do human beings live in a multiverse, or a multitude of subuniverses, phenomena are multiversal as well. A human being is at once a physical entity, a biological entity, a cultural entity, or perhaps an angry god in a play. A book may function just as well as a paper-weight, a weapon, a present, a status symbol on a manager's desk, or an element of décor as it sits on a shelf in a suburbanite's den. It is necessary to note that the activity of framing can misinterpret, fabricate, and even knowingly misrepresent phenomena. However, since framing involves interpretation, and interpretation is based in a descriptive and hermeneutic activity, there is the possibility to return to the phenomenon and to reinterpret from within the same framing modality. Reconfirmed evidence or disappointment is brought forth in intuitive apprehension.

In our discussions, the co-editors expressly have avoided the term, "model." As an illustration of our reason for this, think of a youngster's modeling kit. The reality (plastic airplane or automobile) that (s)he is to bring into being has already been "preconstructed." The real concretum is highly prefigured by what has been already provided in directed operations. Theoretical modeling involves constructing an explanatory system that suggests certain experimental conditions for the manipulation of reality so that certain appearances are predictable. This operational methodology is highly selective in its interest in reality and predetermines the relevancy of data prior to any apprehension of phenomena. In her discussion of Martin Heidegger and the twentieth-century turn to contextualism, Buttimer states that "he called for a sensitive, caring, patient listening to reality—letting reality reveal itself in its own terms—rather than seeking to grasp reality in the language of preconceived models. Heidegger's critique of Western humanism satirized the inherited separation of epistemology and ontology."[12] The co-editors of this volume resonate with this Heideggerian sentiment, although we do not expressly see our project as Heideggerian, for he focussed on the primordial way of Being-in-the-world, and not the secondary way of knowing. When the word, "modeling" came up in our dialogue, we both had the same reaction to it as inappropriately naming the activity we want to advocate. We settled on the word "framing" because it captures a sense of contextualization and because it connotes a knowing process of revealing organized phenomena.

"Model" seems to be a term that is best suited for explanatory science, especially for problem solving. The model is constructed from observation, but not on the basis of presenting that which has been observed in its Being, but rather to fashion a set of conditions that adequately explains an appearance that has been observed. The model is constructed for its predictive value, which is proven through testing. For a practical example, it is observed that a certain type of automobile does not meet safety standards in side collisions. Explanations for the flaw lead to the creation of a model. It is predicted of the model that it will meet safety standards. This model is to be tested. The usefulness of this modality of modeling is obvious, but if employed exclusively can lead to narrow operationalism exhibiting the dangerous tendency to force reality into patterns of utility with unforeseeable consequences. This *way of modeling* historically links up with a mechanistic point of view.

In geography, "modeling" can be interpreted as appropriate to the transcendent type of formism where structural explanations are sought for the patterns of space and society.[13] Phenomena are to be observed and explained from the standpoint of structures not immediately apparent. A geographer applies a structure (model) or articulates a modeling theory that can supply an explanation. Marxist geographers sometimes work in this way. For example, in *Spaces of Hope*, David Harvey provides an explanation of globalization by uncovering the "hidden" (transcending the surface manifestation) principles and processes that have allowed it to emerge as an observable phenomenon.[14] Golledge and Stimson have created the typology of model types and have enumerated models subsumed under those types for the geographical study of decision making and choice. The creation of the typology is itself "formistic." Discriminatory factors in their typology of models are scale (aggregate versus disaggregate, e.g., an individual unit) and category (normative, descriptive, behavioral, and applied).[15]

Philosophically, "modeling" is suggestive of neo-Kantian constructivism.[16] The neo-Kantian view is that all knowledge supplies cognitive form to an inchoate (formless) reality. A knower constructs the formal a priori categories of a type of science (beyond Kant's twelve fundamental categories) in order to make the phenomenon intelligible. We reject the form and content distinction upon which this constructivism is based. Our position removes the tension between the so-called immanent and transcendent formism, which is created by the form and content dichotomy. The source for our position is Edmund Husserl's doctrine of categorial intuition that is found in his sixth logical investigation.[17] Martin Heidegger and Maurice Merleau-Ponty, as well as other major phenomenologists, cite the uncovering of categorial intuition in the sixth logical investigation as that fundamental breakthrough which makes phenomenological inquiry possible.[18] The mind constitutes evidence on the basis of categorial intuition, which opens the possibility of a material a priori science—phenomenology. The content does not depend upon the categories of the mind to supply the organizing form—Kant's mistaken faculty psychologism according to Husserl. In terms of our present concern, this means that for us "transcendent formism" need not be a formal abstraction (Plato), nor a construction (Kant), nor the hidden principles of explanation (Marxist methodology). Alleged "transcendent structures or forms" just as those of immanent formism are apprehended in the acts of meaning-constitution and fulfilment. And, on the basis of Husserl's categorial intuition, this use of the terms "immanent" and "transcendent" are no longer based on preemptive epistemology. Higher level organizations, complex structurings are given intuitively to higher level categorial acts. On the basis of empirical epistemologies all the work of synthesis has to do with mental habits that organizes sense data. On the basis of rationalism, forms are objectively transcendent, reached only by pure thought, and do not arise within appearances. With Kant, all structurings are the formal categories of the mind. But with Husserl's widened empiricism, phenomena are "imbued with syntax"—parts and wholes relations are intuitively given. What was once given over to theory, construction, or explanation, due to biases inherent to epistemological theories, are to be handled through description, for a radically descriptive science involves evidence based on intuitive apprehensions more sophisticated than hitherto had been thought possible. This is not to disregard the transcendent formism defined by Buttimer, that is still a *way of formism*. But it is to say that what had been for so long handled through that way can be handled in an expansion of the *way of immanent formism*, an expansion of what is given to intuition that legitimately encroaches into transcendent formism's traditionally held domain.

Unfortunately, geographers tend to think of phenomenology in a limited way—as a field only concerned with descriptions of lived-experience, but Husserl early in his career opened the possibility for an eidetic science that remains object focussed. A material a priori science is an eidetic science—the apprehension of necessary structures or principles given to eidetic intuition. *Eide* are based in categorial intuition. And now we can give a technical explication of what we mean by framing. As already promoted our sense of framing preserves the subject-object unity, for the way or noetic activity of framing uncovers a frame or the noematic way a phenomenon appears through the noetic act. The frame is not constructed or modeled in advance so that reality is molded to it. Rather, the act of framing and the frame-object is an achievement in securing evidence; an act of framing involves the apprehension of a frame. Defined more accurately, framing is a species of categorial activity that seeks to apprehend the organizing principles of the given phenomena. Founded on acts of perception (which apprehends perceived objects), the simple rays of attention geared towards straightforward noematic contents, the higher level categorial act of framing apprehends an organizational aspect of the phenomena as its intentional aim.

As a descriptive enterprise, framing reveals a frame-structure that manifests as a function of the knowing-activity of the person apprehending it. Seeking an explanation too soon prempts the descriptive enterprise and the apprehension of categorial evidence. The earlier theories of "transcendent formism" then circumvent what could have been gained without constructive or speculative biases or operational manipulations. And so framing offers a descriptive relational hermeneutic, that is, it appears through gathering the sedimentations in the acquisition of knowledge. It is not merely interpreting wholes on the basis of its parts, or parts in relation to a whole. It is the parts/whole organizational relation that is the frame, which is summoned by a perspective that has built up in the knowing activity. In other words, the knowing activity is a process of building a meaning-context that organizes through the act of framing. But this is not to propose a theory or explanation. The framing project involves gathering evidence for categorial intuition. By recognizing that framings are already delimitations of the ways phenomenon display organization, we advocate successive framings that continually apprehend the multiple ways that phenomena present organization. When one is asked to re-frame a question in ordinary parlance, we putatively understand the request to be asking us to reorganize our words so that the contents take on a different value on the basis of a different organization.

The simple judgment, "The manuscript is on the table," presents a categorial object that can be apprehended in a single glance. The "act" of framing is not meant to be a single act but a complex of acts by which organization emerges. But the emergence of the organization is not yet a frame. This is because a frame is not merely an objective category. A frame requires the act of framing to constitute it as such.[19] And so through the sedimentations in the acquisition of knowledge the phenomenon's organizational frame emerges, and this frame is constituted through the act-achievement of framing. Yet, let us return to our categorial in order to illustrate in a simplistic manner an act of framing. "The manuscript is on the table" is based on a parts/whole relation manifesting in the field of perception that is articulated in the judgment. Obviously a parts/whole relation is an organization, but the organization is not yet made thematic as a frame in an act of framing. An act of framing would involve recognizing "*that* the manuscript is on the table" is the organizing structure of interpreting the multitude of phenomena given to perception. Other organizing features may be present and other organizational features could become thematic in another act of framing. Nevertheless, in the thematic grasping of an organizational aspect of the phenomena, an act of framing is filled by intuitive evidence.

The *framing of geographical meanings* indicates that the organizational moment in the phenomenon apprehended concerns its spatial structurings. "Meanings" necessarily has to do with human significance. So, what the frame exhibits is constituted within the framing activity. We do not include in the title of this volume the words, "human meanings," we use, "geographical meanings." In an earlier volume we wrote, "Human expression is a spatial expression; it exudes its *geographicity*."[20] All meaningful human events exhibit a spatial component and so all lived-experience exhibits geographicity—qualitative structurings and processes of gesturing meanings. Merely think of what it is like to ride a bicycle. It is likely that you experience somatological ghost gestures flowing through your body. These activities of your nervous system are at once spatial manifestations, even while you are merely sitting in your chair. All thought is embodied thought and thus it necessarily has a spatial component. The spatiality of the body schema, even the ghost gestures running through your musculature for speech while you read this sentence, involves geographicity. Geographicity, or, literally, the character of earth writing, implies a qualitative significance in topographical manifestations. Even the forming of sounds as you speak exhibits a topographical component in the morphological apparatus activated for speaking. Human beings are earth bodies; humans too are topographical.

Let us return to "geographical meanings" and define it as the spatial moment that is a necessary structurization in all human interpretation. In framing, we use the term interpretation to emphasize the sedimentations in the acquisition of knowledge, the development of a meaning-context, which is necessary for an organization to show forth. Thus "framing geographical meanings" concerns apprehending structural organizations (frames) of the spatial component in all interpreted phenomena. Unless we are interested in framing geographical meanings (engaging in a categorial act of interpreting the organizational moment of the spatial component of phenomena), the frames remain unsummoned and absent to the reflected awareness that we call knowledge. Phenomena can show themselves in various ways, manifesting in various subuniverses of human involvement, and ways other than framing will conceal this aspect of phenomena.

It might be said that framing geographical meanings is the business of geographers, it is merely what they do, and so what is the point of continuing this discourse? It might be claimed by positivists that this discussion characterizes in an obfuscating, esoteric fashion the mundane functions of a particular science. In response, we have elaborated framing geographical meanings in this way, not simply to satisfy the epistemological and ontological interests of philosophers of geography, but precisely to open the activity of *framing geographical meanings* to anyone in any discipline. Framing geographical meanings is not to be taken as the methodological property of geographers. For example, a historian could be studying a particular phenomenon of historical interest. Through gathering specific historical evidence, he is able to apprehend a frame, a meaningful organization of the historical phenomena that allows for him to elaborate a historical interpretation. But by switching frames, that is, taking up interest in framing the geographical organization that can be summoned after his historical categorial activity and intuition, the spatial organization of the historical phenomenon can be apprehended. The initial framing was historical, which involves the historian practicing his own discipline. It is likely that a geographer would not be concerned with that historical frame, or if the geographer had interest in the contents, the geographer would not have apprehended it from the historical framing activities. Now, after apprehending the historical frame of the phenomenon, the historian can make a change in gestalt—he enters another subuniverse—and he inquires after the historically framed phenomenon's geographicity. At this juncture, the historian enters into interdisciplinary space. The historical framing or apprehension of the historical organization of phenomena founds a subsequent geographical framing of the already historically framed phenomena, which constitutes the spatial organizing component within the historical frame.

The Goals of Our Interdisciplinary Form of Study

The Society for Philosophy and Geography takes interdisciplinary study seriously, for we recognize that lacunae in knowledge are a function of partitioned delineations by which the research that results in a framing is not subject to further ways of framings across disciplines. Further framings present the phenomenon in other of its organizational modalities. Interdisciplinary study, as we have conceived it here, takes advantage of research already accomplished. By applying the framing activities of our own disciplines we are able to apprehend an organizational moment of a phenomenon. Others in one's own discipline would have a professional interest in our framing, but it likely remains unknown to those in other disciplines. The purpose of our society is to support a new breed of interdisciplinarians who are willing to enter a *between space* and specifically to make a geographical turn concerning their research. In this way the study of a particular frame is then taken up from the standpoint of a geographical way of knowing, one that would have never been brought to "the light of day." There is a geographical moment to all phenomena and thus a geographical turn can always be made.

Thus, there is a creative dimension to our framings. The creative moment involves our methodological way, which creates a clearing for geographical frames to appear. By becoming attuned to the geographical moment of experience, by becoming sensitive to geographicity, our researchers have been able to frame those structurings and processes that have appeared in the clearing.

We have arrived at our sense of framing through a philosophical grounding of the concept. And, we have taken a philosophical position in articulating it. Thus it could be objected that we have been coercive or have heavy-handedly fitted the essays into our orientation. But our editing methodology was actually a form of apprehending a frame. When we collected papers for possible publication, we had the goal to create a coherent volume and not just have a collection of essays. This is difficult when most of the papers collected come from a conference with a general theme. We were looking for threads within the papers that could give us an abiding coherence. It became evident to us as we discussed the contents of the papers that within their turn toward the spatial component of the topics, a structure was either presupposed, indicated but non-thematic, or had comprised the thematic concern. This structure, then, appeared in the content of the research through the turn toward geographicity, which is the general sense of the goals of the society's research. As co-editors we recognized these structures and for us they became thematic, for they not only seemed to supply the thread of coherence for the volume, but also provided evidence that our interdisciplinary project was exhibiting success. We contacted the authors and asked them to make the framing of the geographic meanings the theme of their essays, rather than a bi-product of their original concerns. Thus the structure was already apprehended in their interdisciplinary turn, but had yet been constituted as a frame through the framing activity. This task the authors accomplished upon subsequent development of their research and revisions in their essays. Even though we have provided a philosophical articulation of the orientational significance of this volume, we as co-editors played the role of directors and producers, and did not supply the scripts.[21] The results compiled in this volume reflect the ideal of the society and the conferences—bringing the various disciplines together to create new *ways*.

Overview of the Volume

This volume is divided into three sections. Section One (chapters 1, 2, 3): Framing Historical Contexts; section Two (chapters 4, 5, 6): Framings Substantive Theories; section Three (chapters 7, 8, 9, 10): Framing Case Studies of Specific Time-Places.

In the first chapter, "Herodotus and the Origins of Geography: The Strange, the Familiar, and the Earthbound," Dennis E. Skocz frames the geographical component of Herodotus' writings, which he characterizes as the way of the aerial and idiographic. He illustrates Herodotus' method of discovery in contrast to the apriorism of the mapmakers of his day. Herodotus provides "ethnographies" from the standpoint of geographies—the grounding of cultures in place. Herodotus is attentive to the differences in cultural geographies. Skocz describes Herodotus' approach to understanding the other, which includes three techniques: pairing, reciprocity, and analogue. The way that Skocz presents Herodotus' geography corroborates what we have called the immanent formism of the Greeks. But as Skocz takes up themes concerning the other's religion, morality, sense of justice and hospitality, Skocz rightly senses a shift in Herodotus' voice, which takes us beyond the immanent formism to which Herodotus is committed and into questions concerning the transcendent formism of which he was critical. And thus with Herodotus we witness what Buttimer has indicated as the problematic tension in Greek geography between giving both the historical elements and transhistorical structures their due and, as important, their relationship.

In the second chapter, "Conceptualizing World Environmental History: The Contribution of Immanuel Wallerstein," Jason W. Moore's scholarship reframes Wallerstein's *The Modern World-System I* as a volume that is strong in ecohistorical content, especially highlighting the historico-geographical modifications that occurred during the transition from feudalism to capitalism. Thus, Moore argues that Wallerstein's theoretics can be framed as a world environmental history in addition to its framing as an economic history. Moore shows how Wallerstein pays careful attention to the production of space and thus economic activity is always interpreted as embedded in, and transformative of, landscapes. Moore shows that Wallerstein emphasizes the interconnection between the quality of the transformations of landscapes and the quality of social relations, which truly constitutes an environmental history. We have placed Moore's chapter before Kirkman's on Rousseau because the historical content is prior to Kirkman's historical and contemporary discussion concerning Rousseau.

Robert Kirkman's "Rousseau in the Suburbs: Geography, Environment, and the Philosophical Tradition," is a study of various incompatible framings of Rousseau's thought. Kirkman injects a healthy contextualism into the practice of framing. For as we have said, the framing of geographic meanings is an exercise in interpretation, and thus it entails the parts/whole problematic. Kirkman cautions that framings can be manifest through the process of narrow selection that neatly forms the frame, but if taken for the whole, it can be quite aberrant.

Kirkman follows a methodology of collecting the framings, contextualizing them. He then observes their reflection in the inconsistency of values that are exhibited in the contemporary landscape. Kirkman begins by framing Rousseau as a suburbanite, as someone who prefers the "middle landscape" between wilderness and the city, far enough from the city's corruption, but close enough to enjoy its amenities. Rousseau is seen as upholding the values of leisure and privacy inscribed in the idea of suburban living. Rousseau also contributes to the romantic sentiment of anti-urbanism. Kirkman explains that this framing is based on a selective reading of Rousseau as well as a selective reading of the processes of suburbanism. He then cites environmentalists who find congeniality in Rousseau's thought. But he also cites humanist thinkers who read Rousseau's advocation of freedom as entailing a sharp separation from nature. Against these selective readings, Kirkman looks at those thinkers who note the complexity of Rousseau's thought. He emphasizes the complexities in notions such as "pastoral ideal" and "nature," and suggests that their ambiguities exhibit a seeming dichotomy between environmentalism and suburbanism that is far too simplified. A much more accurate understanding is gained through a contextualized framing by which there is overlap between them that should not be tidily dismissed and should be recognized as the place where thorny environmental issues arise.

In my own chapter, "Toward a Phenomenology of Cognitive Mapping," I frame a genetic phenomenology of cognitive mapping—knowledge of our whereabouts and how we image getting around our environs. I begin with a fundamental structure introduced by Husserl and advanced by several of the major phenomenologists—the lived-body as the zero point in its own system of spatial orientation. I argue that Husserl's grounding of this structure is inadequate for describing an essential aspect of lived-geography. Husserl's methodology at the time of describing the lived-body's coordinate system, static phenomenology as practiced through the *Cartesian way*, makes an unjustified leap into "transcendent formism." Just short of supplying the most crucial structure for the lived-experience of cognitive mapping, Husserl moves to the abstraction of Objective space. I argue that Husserl's "mistake" does not allow for the proper description of the highest level of cognitive mapping—the synthesis of the multifarious orientational systems mapped through the lived-body's relation to its environment in a unifying system that transcends the limitations of the lived-body's constitutive powers. I argue that the Cartesian way must be abandoned for a phenomenology that takes into account the embodied subject and the embodied world as *co-constitutive*. I take a genetic approach in order to show the logical progression in the structures of significance. This genesis pinpoints what the embodied subject achieves through its own constituting powers and then what structures the "embodied world" (Earthbody) contributes in order to achieve a cognitive mapping that adequately accounts for spatial understanding and orientation.

In chapter 5, "The Die is Cast: Boundaries of Time, Boundaries of Space," John M. Rose frames a dynamic geography of human events. Rose claims that as a technologically oriented civilization our preoccupation with conquering, managing, and possessing space has skewed our understanding of boundaries as merely spatial. According to Rose in order to understand the ontological character of boundary in an adequate fashion, human geography must frame the temporal dimensionality of boundaries. Rose illustrates his theory through three examples: the Dionysian ritual in ancient Greece, the myth of Remus and Romulus, and Caeser's crossing of the Rubicon. The most important point concerns the transgressions across spatial boundaries interrelated with the temporal ordering of acts in time, which are irreversible and irrevocable. The boundary can only be understood adequately as a passing of one time into another. Rose returns to Aristotle's theory in order to explicate the act of time cognition. This allows him to discuss "decisional embodiment" in terms of temporal demarcation. The production and enactment of space, which are created as time framings, constitute boundaries.

In chapter 6, "Toward a Geography of Material Artifacts and an Outline for Synergetic Geography," the co-authors, François-Xavier Nzi iyo nsenga and Gary Backhaus, argue for the need to create a definitive geography of material culture. This geography of human artifacts is framed as a component in a threefold geography that also includes physical and human geography. It is further argued that Buckminster Fuller's physics/metaphysics allows for a common vocabulary and treatment such that the earth's energy as the interrelation between these three geographical taxonomies can be examined. A structural outline of the geography of material culture is developed.

In chapter 7, "A Contextualized Science and the Changing Landscapes of India: A Case Study of Science as a Graft," Deepanwita Dasgupta frames a revisionist doctrine of the role of science by advocating its contextualization. Science must relinquish its hegemony and become a partner in a culture's overall worldview. Dasgupta examines the history and culture of India as a case study. She is optimistic that there can be convergence between India's venerable traditions and modern science, but only through a thoroughly contextualized framing in which the leading role of science is highly tempered in a cultural dialogical network. By examining various weaknesses of the traditional hegemony of science—described as utopian, contextless, "Big," laissez-faire—Dasgupta seeks to show that the sense of science can only be understood through its external relations. She illustrates "context sensitive science" through a discussion of two models of agriculture that had been adopted in India. In another section she takes up the traditional metaphysical views of India and seeks to modify them in a way that can guide scientific contextualization.

Lawrence A. Peskin draws attention to the notion of the geographical dimension of knowledge-acquisition in chapter 8, "Pirates and the Geography of Knowledge: America and Algiers in the Late-Eighteenth Century." In his case study of the Algerian crisis of the late 1700s, Peskin describes the qualities of various mediums for the transmission of news events, oral comunication, letters, official documents, and news reports and how such transmissions shaped a Public Sphere and individual and group identities. He also discusses the cause of an expanding geography of knowledge, which in this era of merchant capitalism concerned the needs of commerce. This theme obviously shaped the contents of the news, including what is constituted as geographically important. Thus, an "informed citizenry" required a great deal more global information. At the same time that cosmopolitanism was developing through the globalization, nationalism was rising as well. So, the geography of knowledge provides important clues for political geography. By framing the significance of the geography of knowledge, the mapping of knowledge dissemination, Peskin provides key insights that are valuable to many areas of concern.

In chapter 9, "Finding the There There: Local Space, Global Ritual, and Early Cold War Berlin," Paul Steege provides a very fine example of an interdisciplinary framing. He emphasizes the geographical component of historical framing through his case study of the city of Berlin during the Cold War. As macrohistorical narratives have given way to microhistorical approaches, Steege promotes an alternative framing (a third way) that accounts for the tension between symbolic (macroscopic) and material (microscopic) meaning-contexts. Steege argues that this approach will place much emphasis on the geography of an historical event, and thus it is inadequate for historical analysis to treat place as merely the setting and name for historical events. Temporality and causality have largely characterized historical events at the expense of the meaning-producing role of the particular contexts, the locations, through which material significances are concretized. Symbolic meanings created by political powers can be quite detached from particular and historical locations. Their symbolic constructions of the meaning of a contested space may not be reflected in the actual meaning-context of the locality. Thus the constructions of these two respective meaning-productions is best understood by framing the geographical location or site of their tensional encounters. Steege argues that the tension between these two levels of meaning—the great powers' decision-makings and the material struggles on the ground—is truly constitutive of the Cold War. Geography, then, discloses the spatial reality of the tension, which means that the geographical nature of the events is necessary and intrinsic to historical interpretation; geography is not merely a backdrop, rather it is co-constitutive. In Steege's framing, history works along with geography as necessary to its own projects. As Steege has built a bridge towards geographers, perhaps geographers could collaborate with him in a geohistorical framing of two levels of meaning in tension.

Notes

1. Anne Buttimer, *Geography and the Human Spirit* (Baltimore: The Johns Hopkins University Press, 1993), 89.
2. Buttimer, *Geography*, 89.
3. Steven C. Pepper, *World Hypotheses* (Berkeley: University of California Press, 1942).

4. Buttimer, *Geography*, 93-94.
5. Buttimer, *Geography*, 95.
6. Buttimer, *Geography*, 95-96.
7. John Kirtland Wright, *Human Nature in Geography: Fourteen Papers, 1925-1965* (Cambridge, Mass.: Harvard University Press, 1966).
8. Buttimer, *Geography*, 79.
9. Buttimer, *Geography*, 81.
10. Thomas S. Kuhn, *The Structures of Scientific Revolutions* (Chicago: University of Chicago Press, 1970). I would suggest the chapter, "Revolutions as Changes of World View," for Kuhn discusses the revolution of Galilean mechanics, which replaces Aristotelian formism as the dominant orientation.
11. Erving Goffman, *Frame Analysis: An Essay on the Organization of Experience* (Boston: Northeastern University Press, 1986). We have drawn insights from Goffman's definition of frame analysis: "the examination . . . or the organization of experience," 11.
12. Buttimer, *Geography*, 68.
13. Buttimer, *Geography*, 115.
14. David Harvey, *Spaces of Hope* (Berkeley: University of California Press, 2000). See the chapter, "Contemporary globalization."
15. Reginald G. Golledge and Robert J. Stimpson, *Spatial Behavior: A Geographic Perspective* (London: Guilford, 1997), 48.
16. For a good introduction to neo-Kantianism, see Gary Backhaus, "The Neo-Kantian Predicament," in *From Kant to Weber: Freedom and Culture in Classical German Social Theory*, eds. Thomas M. Powers and Paul Kamolnick (Malabar, Fla.: Krieger Publishing Company, 1999).
17. Edmund Husserl, *Logical Investigations Volume II*, trans. J. N. Findlay (London: Routledge & Kegan Paul, 1970).
18. Martin Heidegger, *Being and Time*, trans. John Macquarrie and Edward Robinson (New York: Harper & Row, 1962). "But to disclose the *a priori* is not to make an '*a-prioristic*' construction. Edmund Husserl has not only enabled us to understand once more the meaning of any genuine philosophical empiricism; he has also given us the necessary tools. '*A-priorism*' is the method of every scientific philosophy that understands itself. There is nothing constructivistic about it," 490. Maurice Merleau-Ponty, *Phenomenology of Perception*, trans. Colin Smith (London: Routledge & Kegan Paul, 1962). "Phenomenology . . . puts essences back into existence, and does not expect to arrive at an understanding of man and the world from any starting point other than that of their 'facticity,'" vii.
19. See Edmund Husserl, *Ideas Pertaining to a Pure Phenomenology and to a Phenomenological Philosophy Second Book*, translated by Richard Rojcewicz and André Schuwer (Dordrecht: Kluwer Academic Press, 1989). What we call framing is a species of the theoretical attitude. "'*Categorial*' objectivities thus are constituted (in a quite definite sense: objectivities of thought), which, however, for their part, first become *theoretical Objects* precisely when the theoretical subject intentionally focusses on these new objectivities (i.e., above all, on states of affairs, collections, etc.) and so performs new acts which grasp them in their being and determine them theoretically; thus these acts are subject-acts, predicate-acts, etc., of a higher level," 7. So, the framing activity is a theoretical act that raises the frame-object from straightforward apprehension (no matter how complex) to an "enframed" status—an object presented in the modality of knowledge.
20. Gary Backhaus and John Murungi, eds. *Transformations of Urban and Suburban Landscapes: Perspectives from Philosophy, Geography, and Architecture* (Lanham, Md.: Lexington Books, 2002), 2.

21. Although this introduction is written in my voice, the development of the conceptualization of framing involved the ongoing conversations with my co-editor, John Murungi.

Chapter 1

Herodotus and the Origins of Geography: The Strange, the Familiar, and the Earthbound

Dennis E. Skocz

Tradition calls Herodotus the "father of history," but very arguably he has equal claim to the title "father of geography." His inquiry, *historein*, is marked by an infectious and insatiable curiosity about the other. Herodotan geography—a situated, ecumenical inquiry into the "many worlded" world of his time—would seem to have special pertinence and utility for our own world—transformed by unrelenting globalization and far-flung diaspora.

This inquiry aims to show how Herodotan geography enables us to understand the world of the other in terms of its earthbound locality. Recognizing the potential for conflict in cross-cultural encounter in our own time, it examines as well how Herodotus negotiates the difference between the strange and the familiar, not only to understand but also to deal with the other.

Herodotan Geography: Lived Space and "Areality"

For Herodotus geography begins in travel. Classicist Edith Hamilton called Herodotus "the first sight-seer in the world."[1] According to traditional scholarship, "at the most moderate estimate, his travels covered a space of thirty one degrees of longitude, or 1700 miles, and twenty four of latitude, or nearly the same distance." The famous eleventh *Brittanica* tells us:

> He traversed Asia Minor and European Greece probably more than once; he visited all the most important islands of the Archipelago-Rhodes, Cyprus, Delos, Paros, Thasos, Samothrace, Crete, Samos, Cythera and Aegina . . . Sardis . . . the

> Persian capital Susa, . . . Babylon, Colchis, and the western shores of the Black Sea as far as the estuary of the Dnieper; he travelled in Scythia and in Thrace, visited Zante and Magna Graecia, and explored the antiquities of Tyre, coasted along the shores of Palestine . . . and made a long stay in Egypt.[2]

Some scholars accuse Herodotus of inventing most or all of his travels but, according to contemporary scholar James Romm, these scholars are a small minority and the countries which Herodotus claims to have visited would not have been "terribly difficult for a well-to-do fifth century Hellene to visit."[3]

Travel launches Herodotus on his exploration. It inserts him into the physical settings that become fields for exploration. Travel also puts him into proximity to the other, the foreign, and makes possible his inquiry into the relationship between a people and its land. For Herodotus the earth is never an abstraction and is nearly always considered a human habitat, or rather the setting of numerous such habitats, each to be taken on its own terms. Disputing the Ionian claim with respect to extent of the territory of Egypt, Herodotus declares, "Egypt, I consider, is the whole extent of territory *inhabited* by Egyptians, just as Cilicia is the country occupied by Cilicians or Assyria the country occupied by Assyrians."[4] [Emphasis added.]

First-hand experience, his own or that of other eyewitnesses, is the foundation for understanding the earth and its peoples. One scholar counts over a thousand uses of the pronoun "I" by Herodotus.[5] The usage does not reflect "ego." On occasion it signals an opinion on its way, but usually it attests to his having observed for himself or the effort to obtain first-person experience of something. Herodotus the geographer is on scene but keeps out of the picture. Regarding the Nile and the great changes it works on the landscape, Herodotus writes,

> *I have observed for myself* that Egypt at the Nile Delta projects into the sea beyond the coast on either side; *I have seen* shells on the hills and *noticed* how salt exudes from the soil to such an extent that it affects even the pyramids; *I have noticed*, too, that the only hill where there is sand is the hill above Memphis, and—a further point—that the soil of Egypt *does not resemble* that of the neighboring country of Arabia, or of Libya, or even of Syria, but is black and friable.[6] [Emphasis added.]

Travel, then, is essential to the Herodotan geographical project. For this kind of geography, travel can be taken as translation in two senses.

Translation is first movement from one place to another. "Translate" derives from the Latin *translatus*, the past participle of *transfere,* which means to carry or bear across. The first definition of "translate" is: "to bear, remove, or change from one place or condition to another."[7] As movement, travel unfolds in time. Distances in space are measured by the times needed for a traveler to traverse them. Thus, for Herodotus, the greatest breadth of Egypt is a "two months' journey." Or again, he writes, "Above Heliopolis, for a distance of a four days' voyage up the river, Egypt is narrow." One more example: "From Heliopolis to Thebes is a nine days' voyage up the Nile."[8] The earth as scene of travel is *not given all at once* in its entirety but rather *landscape by landscape*. Herodotus's account builds as a narrative. His geographic exposition proceeds from one place to another contiguous to it or next in the circuit of a trip. Libyan tribes of the "sandbelt" are thus described in sequence starting from Egypt and *going* west.[9] The temporality of his inquiry into places points to the fact that his geography is a situated investigation. It draws on the lived experience of what are lived spaces, on what unfolds and presents itself to observation over time. Such an investigation can offer an integral description of each area in its distinctiveness but is ever incomplete and open-ended.

In the 1930s American geographers debated the nature of their discipline and Richard Hartshorne, a key interlocutor, argued that geography distinguishes itself from geophysics and earth science by its "areal" (i.e., area-oriented) approach to the earth.

Hartshorne writes:

> The existence of geography as a field of study rests on the general human interest in the *different* character of *different parts* of the world. . . . This [geography] asks us to describe and explain . . . *all* the features of a *particular* part of the world which are *distinctive* of that part in contrast with others. Geography has, therefore, imposed upon it from the start, the difficult—in a complete sense one may say impossible—task of integrating many kinds of phenomena found in *areas*.[10] [Except for "all," emphasis added.]

By practice, Herodotus is an "arealist." His areality and insistence on firsthand knowledge or lived experience of earth and other immunizes him—not completely but substantially—from the *apriorism* of a Hegel, for whom the character of the separate continents and their respective races was deduced from the "general planetary life of the nature-governed mind."[11] For Hegel the differences among the continents were necessary and not contingent, and the differences among races "go together with these [continental] differences."[12] Hegel parses continents as though they were concepts. There is first the difference between the Old World and the New World. The New World "exhibits only the general difference of the north and the south" whereas the Old World is "sundered into specific diffences," viz., the three continents of Asia, Africa, and Europe. These three form a neat dialectic. Africa represents "undifferentiated unity"; Asia, "unmediated antithesis." Europe incorporates the best of both and "reveals the unity" of the other two continents of the Old World, since "mountain and valley are not juxtaposed as two great halves of the continent as in Asia, but everywhere penetrate each other."[13] Thus topographical features serve as thesis and antithesis for the purposes of sustaining a system. Local geography is made to fit a universal conceptual scheme.

The upshot of this dis-located approach—the knower (Hegel) stands outside the place known, has no immediate knowledge of it—is a geography which ignores, dismisses, and simplifies the physical and cultural diversity of the earth. In Hegel's case, it leads to a blatant racism as peoples and their regions are ranked within his geo-cultural order.[14] Hegel's racism and its relation to his geography is easily the subject of a separate reflection, but its underlying logic is adumbrated in a passage where Hegel relates what he takes to be a "dormant mentality" characterizing the African to the "compact, differenceless mass of the African continent."[15] For Hegel, differentiation and mentality go hand in hand. Greater differentiation—more precisely diffentiation mediated and subsumed within a higher unity—marks a higher degree of mentality or spirituality. Because the European mind "opposes the world to itself, makes itself free of it, but in turn annuls this opposition, takes its Other, the manifold, back into itself into its unitary nature," it is, for Hegel, nothing less than "self-conscious Reason" itself.[16]

For Herodotus the earth is not an ideal whole whose parts (continents) express essential moments or aspects of that whole. Herodotus derides the "a priori geography" of many mapmakers of his day. "I cannot help laughing at the absurdity of all the map-makers—there are plenty of them—who show Ocean running like a river round a perfectly circular earth, with Asia and Europe of the same size." Herodotus proceeds to describe the location and size of the continents, reasoning from what was known at the time and referring at length to reported explorations to circumnavigate Africa.[17] Concerning the latter, the unsuccessful attempt of a certain Sataspes, commissioned by the Persians, is more believable than the Phoenician-manned party sent by Egyptian King Neco. Even if both accounts are bogus, however, it is significant that Herodotus refers to what reported explorers have to say about the size and shape of landmasses. Empirical knowledge, and not a tradition that held for map symmetry, is important. For Herodotus, one could speak knowledgeably about Asia since "the greater part of Asia was discovered by Darius; . . . with Europe, however, the case is different; for no one has ever determined whether or not there is sea either to the east or north of it."[18] Again, the point is not whether Herodotus had correct knowledge or even whether he drew valid inferences from data. In these cases the relevant fact is that he looked to what could be known by discovery to determine geographical truth.

Herodotan geography is painstakingly pieced together over time from the in situ experiences of one who travels, who traverses the earth and who traces its contours. Herodotus's method is discovery, not deduction. Differences are not fodder for a general theory; they make up a diversity that resists categorization or subsumption in a dialectic. Herodotus sometimes compares cultures (Egyptians and the "rest of mankind," Greeks and Persians) and frequently relates peoples to their local geographies. His observations, however, are not adduced as evidence for a law or principle, nor do they establish ranking or hierarchy.

The Familiar and the Strange

Travel, as translation or movement across the earth over time, as exploration, is critical to geography as Herodotus practiced it. But, translation has another sense. Translation is also interpretation, of one idiom into another. This too is central to the Herodotan undertaking.

Travel, of the kind that Herodotus describes, is a translation or movement from the familiar into the strange. Travel makes one a stranger in a strange place. It both necessitates and makes possible interpretation. The transposition from the familiar into the strange can yield knowledge of the strange, but the traveler must avoid two extremes.

In one case, the strange is taken as the exotic. The strange remains strange and is prized as such. The traveler would rather delight in the bizarre than come to understand what is other. In this posture, the stranger never understands that he/she is the stranger or "other of the other" and that what is strange for him/her is the familiar of the other. The traveler becomes the collector of curiosities, the spinner of tall tales. The transposition into a new and strange world does not become the occasion for knowledge. This is because the traveler enjoys the aura of strangeness and prefers to remain in aesthetic suspension at the margins of the other's world rather than risk losing the "feel" of the exotic by crossing the threshold of the other's world with the intent of understanding it. Herodotus often falls prey to this—or seems to. On a first reading, one is struck by the Herodotan bestiary: the flying snakes of Arabia,[19] the sheep with tails 4 1/2 feet long (for whom the shepherds make carts fixed to the tails so they do not drag) or sheep whose tails are 18 inches wide,[20] and the ants as big as foxes.[21] In Western Libya, Herodotus relays reports of "horned asses, dog-headed men, headless men with eyes in their breasts (I don't vouch for this, but merely repeat what the Libyans say), and small snakes with a single horn."[22]

The other extreme is going native. One assimilates the lifestyle of the other, becomes absorbed in the world of the other so that it becomes fully familiar. There is nothing objectionable or wrong in this, but the assimilation into the world of the other, such that the strange becomes utterly familiar, relinquishes the possibility of exploiting one's strangeness in a way the renders the familiar intelligible.

The familiar is taken for granted. Indeed, it *is* the taken-for-granted. Being taken for granted defines the familiar, and the familiar constitutes the context in which we live. As such, it is not that *about which* we think but the implicit understanding *within which* we think, act, believe, judge, feel—in a word, live. The traveler, as alien, is privileged to know the familiar in a way the inhabitant, as inhabitant, cannot. Herodotus, for example, notices that Persian names which denote magnificence or physical qualities end in the letter *s*. This is a fact of which the Persians themselves are "unaware"; "inquiry," Herodotus says, "will prove this in every case without exception."[23] For the alien, "the familiar" is not familiar; thus, it is not taken for granted: it can become questionable, and its questionability can provoke inquiry leading to a knowledge foreclosed to that taking-for-granted of the familiar.

Herodotan geography is distinguished by a *historein*, inquiry, a questioning that first sees the taken-for-granted as questionable and pursues it with the aim of giving account for it, explaining its cause. The paradigm case is Herodotus' effort to understand the annual flooding of the Nile. For the inhabitants of Egypt—the "gift of the Nile"—there is nothing at all strange or questionable about the flooding phenomenon. It is part of the setting within which they live. It is familiar, taken for granted. Herodotus is amazed that the Egyptians do not see the flooding as questionable, do not seek to find its cause. "About why the Nile behaves precisely as it does I could get *no information* from the priests or anyone else," Herodotus writes. "*Nobody in Egypt* could give me any *explanation* of this in spite of my constant attempts to find out what was the peculiar property which made the Nile behave in the opposite way to other rivers."[24] [Emphasis added.] It takes the stranger to see the familiar as questionable and to begin an inquiry to render the phenomenon intelligible with respect to its origin and cause. "Certain Greeks," Herodotus continues, "hoping to advertise how clever they are, have tried to account for the flooding of the Nile in three different ways."[25] Strangers, not natives, venture explanations.

Herodotus joins other Greeks in offering an explanation. To be sure, Herodotus is wrong. "To put the whole thing in the fewest words . . . during winter the sun is driven out of his course by storms toward the upper parts of Libya."[26] And, his science strikes us as bizarre, but if we go beyond first impression and understand the "natural philosophy" of his day, we see the workings of a methodical mind, a reasoning that systematically explores alternatives and "tests" hypotheses.[27] His practice offers a marvelous example of the privilege of the stranger to take the familiar of the other as strange (i.e., questionable) and then to render the questionable intelligible.

In this paradigm case, Herodotus explores a natural phenomenon both wondrous and related to the lived world of the Egyptians for whom annual flooding was/is the material basis of their lives.[28] Additionally, Herodotus's foray into geophysics is thoroughly "areal." He is not seeking universal principles of meteorology and using Nile flooding as evidence but just the opposite. With what he already takes to be general principles of meteorology and physics he seeks to give account of a singular phenomenon peculiar to a particular place—a phenomenon which determines the lifeworld of its inhabitants. Herodotan geography is areal—his earth, a habitat for humans.

The Other's Place

The habitat-humanity or earth-to-culture relationship becomes the explicit theme of many Herodotan "ethnographies"—ethnographies that are as much geographies by virtue of their grounding of culture in place, their giving an earthbound account of other lifeworlds.

- Herodotus describes how the natural environment, its flora, and cycles affect everyday life for the Egyptians of the marsh country, who bake loaves from the dried poppy-head of the indigenous lotus blossom and consume papyrus from the marshes or perhaps "nothing but the abundant fish of the area," dried in the sun upon being caught.[29]

- Likewise the Indians who live in marsh-country eat raw fish caught from boats made of reeds that grow in the waters; clothes are made from a rush that that is beaten and then woven into a matting suitable for breast-plates.[30]
- The influence of land, water, and climate extends beyond food and dress. Scythian life, by Herodotus' account, owes as much to the river Borysthenes, as Egyptian life owes to the Nile. It supports "the finest and most abundant pasture," the "best sorts of fish," and "excellent water for drinking."[31] Warfare and, one may infer, a warrior religion derive from the lifestyle which the land supports. "They have been helped in this [escaping destruction by the conquest of others] by the nature of their country," Herodotus writes.[32] Nomads who raise cattle and move on horseback, the Scythians frustrate would-be conquerors for whom they offer no fixed settlements to attack; invaders must reckon with a people who can fight on horseback with bows and arrows and whose swift maneuver tactics undoubtedly benefit from a life on the move. It is no wonder that the Scythians sacrifice to only one god, Ares, the god of war, that the horse is the preferred and most common victim, and, given the lack of wood in their grasslands home, the Scythians do not burn their offerings but strangle their animal victims and boil their flesh over a bone fire.[33]

Outside the "cameo ethnographies" that abound in the text, Herodotus shows his attentiveness to the influence of local geography on the material culture of a people.

- Nile boats, he observes, are built of the acantha wood of the area, "caulked from inside with the indigenous reed papyrus," and powered with aid of sails of papyrus.[34]
- In the marsh country of Egypt a prolific and pungent plant, the Kiki, is sown along the river and shores of lakes so that the fruit can be pressed and boiled down to produce an oil for burning in lamps—"though the smell is unpleasant."[35]
- Herodotus notes that the rainfall of Assyria is very slight—sufficient only to "burst the seed and start the root growing"—and so "artificial irrigation" is used and "the whole country is intersected by dykes."[36]
- Herodotus tells us about the salt block houses of Libya which testify not only to the abundance of the material, both white and purple, but also to the absence of rain in the desert environment, for "if there were [rain], the salt walls would collapse."[37]

In all of the above examples we see the workings of an earthbound understanding of culture. The effect of such geography is to humanize others obliquely. It may not seem so. Indeed, it may seem biological and Darwinian to take such an environmental view of culture. Notably, the first definition of "habitat" is "the place where a plant or animal species naturally lives and grows." According to *Webster's*, *habitare*, the root word, is the initial word in

Latin descriptions of species of flora and fauna in old natural histories.[38] But if the problem we have identified lies in reducing other cultures to curiosities of an alien observer, then situating the other culture within a local geography approaches that other culture on its own terms, terms specific to its situation, including its placial, earthbound setting. Food, shelter, clothing, social practices, and even religious beliefs are seen to have a basis in the local environment. Smelly lamps and houses built of salt, which might otherwise seem exotic and odd, are understood in terms of the local geography. The other is understood as "at home" in his/her world, on his/her own turf, rather than as a colorful oddity in an alien's travelogue. It is well to point out here that Herodotus is not purveying a geographical determinism. More often than not, he describes cultures apart from their relations to local geography, notably, for example, the Persian and Babylonian. Moreover, he is adept at tracing historical as well as geographical/environmental influences as both become relevant to our understanding a people. The Geloni, for example, build temples in the Greek style to gods whom Herodotus calls by Greek names but they are constructed and furnished with statues and altars entirely of wood. The use of wood attests to the fact that "the country here is forest with trees of all sorts." The Greek style and pantheon owe to the fact that the Geloni "were originally Greeks, who, driven out of the seaports along the coast, settled amongst the Budini."[39]

The Other as Analogue

The consideration of culture apart from place permits us to ask how Herodotus handles the one-on-one encounter of cultures—his own Greek culture with the many others he describes? How does Herodotus deal with the other apart from an explicit reference to place?

We may begin with the famous comparison of Egyptians to the rest of mankind in Book Two of *The Histories*. "The Egyptians themselves in their manners and customs seem to have reversed the ordinary practices of mankind," Herodotus says.[40] Even when one reads the passage in full and not an extract of key points, like that below, one notices that the non-Egyptian behaviors, which are said to be common to the rest of humankind, are not mentioned. Stylistically, of course, it would be tedious to do so, since invariably they would be the contraries of what Egyptians do. But the omission of the contrasting ordinary practices of humankind suggests another interpretation worth mentioning: that these are all too familiar (i.e. may be taken for granted) and that perhaps the terms in comparison are really Egypt and Greece rather than Egypt and the rest of humankind. Herodotus, after all, was a Greek author writing for a Greek readership. He could safely assume that Greeks would regard their mores as normal and that he could neglect adducing evidence to establish that. However one construes the precise terms of comparison, the movement is between the explicitly strange and the implicitly familiar. By comparison with the rest of mankind here is how the Egyptians differ:

- Women attend market . . . men stay at home . . .
- Men in Egypt carry loads on their heads, women on their shoulders . . .
- Women pass water standing up, men sitting down. . . .
- To ease themselves they go indoors, but eat outside in the streets . . .
- No woman holds priestly office . . . only men are priests . . .
- Elsewhere priests grow their hair long; in Egypt they shave their heads. . . .
- They [Egyptian men] practice circumcision . . . the men of other nations leave their private parts as nature made them. . . .
- In writing or calculating, instead of going, like the Greeks, from left to right, the Egyptian go from right to left . . .

At first blush, the comparison seems unflattering, portraying Egyptians as peculiar, if not odd. Herodotus, however, is not one to denigrate cultures wholesale. For him there are no a priori hierarchies of peoples. At times, Herodotus will speak pejoratively or approvingly of other cultures. He will say of the Greeks, for example, that they "in general have a weakness for inventing stories with no basis in fact."[41] The Egytians "are religious to excess."[42] According to him, Ethiopian men are the tallest, longest-living, and best-looking in the world.[43] His estimates rarely reflect a sweeping judgment about a culture in its entirety but are made in respect to a given practice, a certain aspect of life, or particular norm. One exception is a judgement about the peoples around the Black Sea whom he calls "the most uncivilized nations in the world." Even here, Herodotus immediately proceeds to describe the Scythians, who are predominant in the region, with appreciation and in detail for the way in which they manage "their own preservation."[44] Very often it is not clear whether a quality has a positive or negative connotation. In such cases, it is not unreasonable to assume that it has neither and that Herodotus speaks matter-of-factly, as he does most of the time.

We shall see now how the comparison of Egyptian culture with others, far from denigrating Egyptian culture, paves the way for cross-cultural understanding. If Herodotus can be said to have a methodology for cross-cultural understanding, it involves three concepts: (a) pairing, (b) reciprocity, and (c) analogy.

The above-mentioned comparison of Egyptians and others illustrates the way pairing works. The pairing of traits with respect to various practices (eating, work, worship, etc.) "equalizes" the cultures compared. Even if the content of a specific comparison made by Herodotus seems to denigrate the Egyptians, the practice of repeatedly comparing one culture to another or to others generally, doing so with respect to discrete aspects of behavior, and including everything from religion to hygiene, can only invite the reader to consider the differentiation of one behavior vis-à-vis another rather than the practices compared. The result of such a comparing is not a prescriptive ranking of one in terms of the other, but a descriptive regarding of both in terms of their difference. If Egyptian behavior can be judged by the norm of the contrary behavior of others, then why not hold others' behavior to the norm of Egyptian behavior? If Egyptian behavior seems strange from the standpoint of the non-Egyptian, then might not the familiar non-Egyptian (e.g., Greek) behavior seem strange to the Egyptian? Pairing allows for comparing, but comparing works "both ways." Pairing, as described here, opens up the understanding to reciprocity: each is the "other of the other." Reciprocity implies that the other is another self "like me" (obviously not in particulars of behavior but *qua* self) and it allows me to imagine myself as others might (i.e., as an "other"). Colloquially expressed, reciprocity is about putting oneself "in the other's shoes."

Translation as movement—*from* the familiar *into* the strange—leads to translation as interpretation where interpretation is understood via reciprocity: as a *to-and-fro* between self and other.

But what about analogy, our third concept?

Again and again, Herodotus analogizes. He endeavors to understand other cultures by analogy to his own Greek culture. On the face of it, this might seem less than commendable, indeed to be a formula for misunderstanding the other. But, Herodotus and all of the rest of us have at least one good *excuse* and several good *reasons* to embrace analogy—at least a certain kind of analogizing.

Frequently Herodotus describes foreign gods by analogy to the Greek gods. Comparing Greek to Scythian gods, he says, "Hestia is *Tabiti*, Zeus (very properly, in my opinion) *Papaeus*, Earth *Api*, Apollo *Oetosyrus*, Aphrodite *Argimpasa*, Poseidon *Thagimasadas*." [45] Thus the strange is made more familiar. This is not so different from word-for-word translation. How else is one to begin to understand a strange phenomenon except by finding a word that denotes a near equivalent? To criticize the interpreter for using the medium of a familiar language to understand a strange entity or term is to suggest that either we can enjoy an unmediated access to the strange or that we should simply abandon the effort to make the strange, to some extent at least, familiar. If we want to understand the strange, then we must make it familiar with what is already familiar to us. How might Herodotus have better described the strange hippopotamus than by successive analogies?[46] This is the good *excuse* for analogizing.

But Herodotus shows us how to analogize well. Thereby, he gives us several good *reasons* to analogize in service of cross-cultural understanding.

Herodotus analogizes without privileging the analogue. In referring to the Egyptian gods, Herodotus does not assume that the Greek names are prior to the Egyptian. In the order of knowledge, his knowledge, the Greek gods are first (i.e., *known* before the Egyptian). If that were not the case, they would not serve as analogues for understanding the Egyptian. But, in the order of being or in point of historical fact, at least according to Herodotus, the Egyptian names precede the Greek.[47]

For Herodotus, analogy serves as a first approximation to, not the final word for, the phenomenon. Understanding the other is an iterative process. One only begins that process by comparing the strange with something familiar; the comparison is and remains open to correction and precising over time. There is a testing and verification to see if the analogy holds up. Analogizing is more than a one-time, one-on-one comparison. Already in the notion of analogy other terms are implied—A:B :: C:D. Analogy is not just the comparison of two terms, A and B, but rather the comparison of two relations: that of A (the iconography of Io) and B (the iconography of Isis) with that of C (ritual of Io) and D (the ritual of Isis). The utility of a comparison between Io and Isis proves itself when we compare the myths and liturgy of Io with those of Isis. The iconography of both is similar; both are female figures represented with cow horns, but, while Io may receive the sacrifice of cows, Isis may not. The analogy only goes so far. The same is true of the Greek Zeus and his Egyptian counterpart Amun. They are much alike in their core identities and place in pantheon. Yet in pursuing the analogy beyond initial similarities, distinguishing differences emerge. Ritual specifies the sacrifice of sheep to Zeus and goats for Amun, the Theban Zeus; mythology explains why.[48]

These examples from Herodotus' long and painstaking comparison of Greek and Egyptian gods point to another apt use of analogizing: to expose difference as well as identity. The failure of the analogy, Io to Isis, to hold up in the area of liturgical sacrifice serves to underscore what is different and distinctive about Isis and Egyptian religious belief more broadly. Isis rejects the sacrifice of cows; cows are sacred to the Egyptians. Analogizing may be more useful to understanding the other on its own terms when analogy breaks down rather than when it succeeds. The strange is and appears strange because it is not *immediately* familiar to us. It is likely as well that it is not *entirely* like anything we already know. If that is the case, then a process that ultimately leads to defining a difference that spoils any simple coincidence of meaning may be more meaningful for our purposes than one that finds identity at first sight.

Analogy and reciprocity combine in a very illuminating way in reference to the word "barbarian"—which, for the Greeks, originally meant the non-Greek speaking. Early on, if not from the beginning, the word was used to belittle the foreign and keep it at a distance. Herodotus cites the case of the Egyptian pharaoh Necos who refers to the non-Egyptian-speaking as barbarian.[49] The allusion suggests an analogy—Egyptian:barbarian :: Greek:barbarian. It is one that suggests in turn a reciprocal transposition of Egyptian and Greek so that each becomes the barbarian of the other! Herodotus does not draw out the point; it is only implicit in the brief reference to Egyptian usage, but the allusion begs one to make the inference explicit. The reasoning here may seem "overfine," but Herodotus scholar James Romm draws essentially the same inference from the same passage.[50] If we may allow it, the transposition undercuts any *hybris* in the use of "barbarian" as "put down." The relationship between analogy and reciprocity is not peculiar to this example. Analogy, by its nature, would seem to allow for reciprocity between its terms. If analogy and reciprocity work hand in hand and if recriprocal logic levels the playing field for a nonhegemonic or nonreductive relation to the other, then analogy offers not an obstacle but a gateway to cross-cultural understanding.

The Other "Out of Place"

Earlier we noted that the other's earthboundedness, when understood, helped to protect others from becoming exotica in a travelogue. What happens when the other is not encountered on his or her own "turf"? Globalization and diaspora have resettled millions in strange lands, far from home—far from the place that goes to constitute so much of their cultural being. In such a case—that is, the situation of our time—cross-cultural understanding is handicapped. The opportunity of grounding understanding of the other on the other's ground is lacking. The other is and appears dis-located and no longer at the center of the cultural world that has gone to make up his/her being; but at the margins of the world of that self for whom the other "shows up." In such a case, one should expect that at least initially the other will be understood in terms of the categories of the self who encounters the other. Those categories could be specific stereotypes of the group to which the other belongs or a benign (at least, benignly intended) concept of humanity which discounts for individual and cultural differences and accepts each as a equal member of the human race. In either case, the other is seen and understood—again, initially—from the point of view of the self into whose world the other intrudes.

But, the problem today goes beyond the handicap that we have considered so far. If Huntington's thesis is given its due, then ours is a world characterized by a "clash of civilizations."[51] It is not just that one civilization meets another across a gap of ignorance but rather along a divide marked by misunderstanding, resentment, fear, and hostility. In such a clash of civilizations, the religious affiliation/belief of the other appears to be a principal cause of conflict. In such a world, the benign neutrality of the cosmopolitan observer, one who is impelled by curiosity to investigate the diversity of cultures and who regards the norms of human action and behavior relativistically, may fall short of offering what is needed both to understand and to deal with the contemporary situation. It is not the point of this paper that Herodotus' approach is ultimately one of cultural relativism. To be sure, much of his geography, including what we have focussed on, suggests that. We shall see, however, the resources within Herodotan thinking for transcending cultural relativism and addressing the distinctive nature of today's many-worlded world.

Thus far, Herodotus seems to have given us an approach for understanding other cultures based on first-hand lived experience of the other in situ aided by an implicit methodology of pairing, reciprocity, and analogy. None of this needs to be abandoned as we consider the specific challenges presented by displacement, religious fervor, and civilizational conflict. Herodotus's response to these three challenges shall be described respectively in terms of what will be called his (a) idiographic focus, (b) pious agnosticism, and (c) sense of justice and hospitality.

Idiographic Focus

We have already noted that Herodotus eschews a universalizing apriorism that sees each culture as instantiating some law of human nature or some necessary relationship between place and culture. In contrast to the nomothetic approach of the social sciences that seek to uncover correlations between factors like income level and church attendance, Herodotus' geography is idiographic. He approaches each phenomenon in its singularity and wholeness (i.e., each culture holistically and in its own time and place). Herodotus not only avoids Hegel's racism—not a necessary correlate of an a priori perspective, but, in his case, inextricably bound up with it—but also the loss of distinctive difference and individuality that an approach whose aim is universal truth suffers. Herodotus, we have seen, unfolds a truth about culture rooted in locality.

But how useful is such a truth in a world of displaced peoples? There seem to be two ways of answering this question.

First, one who would know an other who has been separated from his/her locality can analogously and imaginatively transport him/herself into the other's world—accomplishing by imagination what Herodotus could do with the first-hand experience he had of the other *in* his/her world. Herodotus' insertion into the earthbound world of the other was immediate and served to check his understanding of the other. To be sure, displacement, as we have described it, denies us the immediacy that Herodotus enjoyed. For the most part, the other comes to us without his/her world and earth-point-of-origin. Our effort to understand must rely on information and imagination. Information here signifies either first-person accounts by the very "others" we would know or first-hand reports by observers of the world of the other. By itself, however, just having information from either source falls short of the mark. Imagination is needed to transform this "data about . . ." into an "as-if-there" re-enactment of the other's lived world. To inform our imaginations, we come back to the necessity of a geographic understanding of the type that Herodotus pioneered. This is a geography ideographic and areal, one that, of course, begins with the earth, but focuses on a region and its singularity rather than the whole earth and its laws. This geography is ecumenical in root sense of the word: earth is home and habitat, or more precisely, *many* homes to *many* cultures. Although we may not be able to undertake the kind of geographic exploration and inquiry that Herodotus demonstrates, we may yet compensate for our own lack of first-hand knowledge of the other in situ through an informed and vicarious transposition into the other's world.

Secondly, even though the other comes to us dislocated, we can apply the essential thrust of local truth to our understanding. What remains of Herodotus's areal and earthbound understanding of the other after one substracts access to the other in situ is an approach which (a) valorizes knowledge that arises from encounter and which (b) takes the other within the context of his/her world. We need not cite the many passages in which Herodotus affirms the value of immediate knowledge of either the lands or peoples he describes. What needs to be added is that even as the other appears outside his/her locality of origin and habitation, the other carries with him/her the world that took form in that locality. The Scythian removed from the steppes over which he/she ranges on horseback would presumably, for a time at least, bear the world of the steppes, one which prizes mobility and a fierce independence. The Geloni, we noted earlier, built Greek-style temples to Greek gods, long after they departed Greek shores to live in forests of the Caucasus with the Budoni. Within the idiographic perspective of Herodotus, our task would be to ground our knowledge of the other in direct encounter and to endeavor to situate the other with the other's world, even if that world becomes detached from its geographic setting.

Pious Agnosticism

"Pious agnosticism" is both neologism and oxymoron. It is meant to describe a Herodotan attitude that seems well suited to understanding and dealing with our world—one where religious fervor often fuels cross-cultural antagonism and enmity. Our reflection will serve to define the expression.

More than once, Herodotus disavows interest in what we might call theological matters. Noting that animals in Egypt are held to be sacred, for example, he avoids any discussion of the reasons for such practices. "Discussion of religious principles," he says, "is a subject I particularly wish to avoid." "Forced . . . by the needs of the story," he goes on to discuss religious practices—"how they actually behave"—but not the "reasons" for the behavior.[52] This is a bit disingenuous, or seems so, given so much discussion of religious belief and practice in *The Histories* and explanations of various kinds put forward by Herodotus. Nonetheless, the citation gives explicit form to a kind of agnosticism, if by that we may mean a reluctance to debate the merits or the "truth value" of certain religious beliefs. This bracketing of the issue of religious truth goes hand in hand with a cultural relativism that seems to come out explicitly at times, for example, in a comparison of Egytian priests and Persian Magi regarding the killing of animals. In contrast to the Egyptian priests who "kill no living creatures except for sacrifice . . . the Magi not only kill anything, except dogs and men, with their own hands . . . they kill them indiscriminantly," Herodotus writes. Then he adds, "Well, it is an ancient custom, so let them keep it."[53] Perhaps more important than such specific expressions of indifference is the relativism that seems to underlie the approach we have elaborated in terms of pairing, reciprocity, and analogy. To be sure, Herodotus is shy of discussing religious principles and his even-handed discussion of cultural mores could be taken as expressing indifference to a truth that goes beyond factual correctness. Nonetheless, there is every reason to qualify these impressions.

We shall see that a kind of piety underlies Herodotan understanding and that it empowers him in relation to others he seeks to understand. Herodotan piety is most evident when he writes of oracles. Oracles play a more important role in the historical narrative of *The Histories* than in the geography, but with respect to the latter, Herodotus says that his opinion regarding the extent of Egypt is "supported by an oracle delivered from the shrine of Ammon."[54] This point suggests that Herodotus credits oracles with truth, although it tells us little about the deeper religious significance of oracles for Herodotus. For this we may turn to the discussion of the famous oracle message that tells Croessus that in attacking the Persians he would destroy a great empire.[55] Croessus was delighted with the information because he assumed that he would destroy the Persian Empire. He learns to his regret that he misinterpreted the oracle and that his attack on the Persians would bring the destruction of his *own* empire. The story of the oracle, its interpretation, and subsequent events becomes a commentary on human hybris and the just desserts of asserting such hybris.

Herodotus does not make an explicit argument for the truth of oracles. For the most part he tells his readers that others credited oracles and like phenomena with truth and took them to have a certain meaning. Herodotus does not tell his reader what he thinks about oracles and the like. It is the prominence that Herodotus gives to the Croessus account and others like it that strongly argues that Herodotus takes oracles seriously and that he understands human action to come up against a divine measure or limit. Herodotus translator Aubrey de Selincourt writes that in spite of the many differences among the humans that Herodotus describes and appreciates, "they [humans] are all dominated by the same unsearchable Fate, which even the gods themselves cannot escape." De Selincourt adds, "It is a primitive philosophy— but it goes deep."[56]

Greek and non-Greek alike have access to the divine via oracles and, we should add, via dreams and omens in natural phenomena. But the messages that originate in these are often ambiguous; they are not clear directives but intimations. The human must interpret their meaning and the response becomes a test of whether the human has taken the proper measure of him/herself. Herein lies the nucleus of a religio-philosophical understanding. If this represents Herodotus's understanding, as it very arguably seems to, then his agnosticism (reluctance to privilege one cultural practice or belief over another or to examine the grounds of belief) must be qualified by a piety whose thrust is to regard human action as ever measured and tested in the interplay of divine intimations and limited human self-understanding.

How does pious agnosticism bear on Herodotus' understanding of the other and, more pointedly, what relevance does it have for us in a world rent by religious divisions?

The answer to the first part of the question is suggested in statement by Schleiermacher made two centuries ago. "Only the person who settles down in such a [determinate] form with his own religion really establishes a firm abode and, might I say, an active citizenship in the religious world."[57] If Herodotus is the pious Greek he so arguably seems to be, then being rooted in the ancient Greek religious understanding gives him access to the religious world as such. For the purposes of understanding others' religions his own religiosity is not a barrier but an advantage which the impious or those simply lacking in religion cannot call upon. It is his religiosity which allows Herodotus to appreciate and convey the peril hanging over the infancy and boyhood of King Cyrus of Persia inasmuch as a dream concurrent with his birth was interpreted by the Magi of Persia to mean that Cyrus would become a political threat to the incumbent Astyages.[58] The impending doom awaiting the Persians in Salamis is intimated by a cloud that drifts over the Strait—a cloud from which a "mysterious divine voice" issues.[59] The point of these "events" for us is that they illustrate how Herodotus' own religious understanding give him access to the lived experience of others—indeed Persian adversaries—in a way that allows the human significance of their situation to unfold for him and for us.

In our own day, pious agnosticism may well describe a useful response to the religious fervor that drives cultures and civilizations toward conflict. Agnosticism—understood here as reluctance to debate issues of theology—would steer us clear of antagonizing discussions with little prospect of resolution. Agnosticism, as defined here, kept Herodotus "on program"; his project was to understand the other and that understanding embraced the other's religious belief and practice, but stopped short of examining religious principles. By itself such agnosticism, however, falls short of the mark. Disengagement on religious matters can be taken as indifference, and indifference as devaluation (your beliefs do not merit consideration); even worse, it may suggest disdain. Piety, rootedness in one's own religious understanding, works positively to signal seriousness regarding religion and offers a commonality among religionists that subtends doctrinal differences. Neither agnosticism nor piety nor the two of them together are a match for fanatacism of any kind. But to the extent that cross-cultural understanding and communication provide some means of moderating conflict and fostering coexistence across civilizational differences, then the Herodotan formula, to which we have given the name "pious agnosticism" seems well suited to make a contribution.

Justice and Hospitality

The story of Paris, Helen, and King Proteus of Egypt, as told by Herodotus, is a parable of hospitality and justice. It suggests that there is a right and wrong that transcends the mores of individual cultures and peoples and a regard for the other that goes beyond the quid pro quo of reciprocity.[60] Paris and Helen are the very same individuals who figure in Homer's epic. They have come to Egypt, forced ashore by a storm. The Egyptians welcome them and the Greek slaves that were on board the ship are given sanctuary—and therewith freedom—at a temple to Heracles where custom dictates such a practice. The slaves tell the priests that Paris has abducted Helen, and word reaches the king who then chastises Paris, keeps Helen, and instead of sentencing Paris to death, banishes him from Egypt. The story gives pause to one who might think that morals and mores are the same for Herodotus or that simple reciprocity governs relations between natives and strangers. The abduction is described as an abominable crime. The turpitude of Paris expands to include seduction-as-betrayal since Helen's husband is a friend of Paris and theft of Helen's husband's goods. On top of all of this, Paris exploits the passion between Helen and himself to make her an accomplice in his crime. Nothing in the story qualifies the deeds as mere violations of local customs. Proteus, for his part, says, "no matter who it is that has committed this crime against his friend, arrest him and send him to me so that I may hear what he can say for himself." These are not behaviors that might be excused depending on the station or nationality of the presumed offender. Anyone, "no matter who," is a criminal if he/she acts as Paris was accused. Paris, for his part, does not question the jurisdiction of Proteus but puts up a defense, albeit ineffectual.

Proteus finds Paris guilty and gives him three days to leave Egypt; after this he will be considered an enemy. This last point is telling. It implies that what Proteus has done toward Paris is not in the nature of an attack on a foe or revenge by someone who himself has been wronged. Proteus, as the expression goes, "has no dog in the fight." In that sense, his action is disinterested. His invocation of rules (albeit his own) which govern such situations and prescribe consequences, including rules with respect to foreign offenders, tell us that Proteus is acting in the name of justice.

Herodotus tells us he is relating what the Egyptian priests told him, but there is every reason to read the account as representing Herodotus' ethical stance.[61] He says he agrees with the Egyptians that Helen was in Egypt, not in Troy, during the Trojan War.[62] More to the point, Herodotus presents an admirable portrait of Proteus, who comes across as generous, deliberative, and just. If the appearance of these qualities in the account owes entirely to what the Egyptian priests told him, Herodotus does not choose to diminish at all the favorable light in which Proteus shines. At the conclusion of the whole account he draws this moral from Homer's *Illiad*, "sin is always visited by condign punishment at the hands of God. That, at least, is my own belief."[63] Herodotus, who has earlier shown his talent for matter-of-fact description of humanity's diversity, now contrasts perfidy (Paris) and virtue (Proteus) unqualified by place-bound standards, neither concealing his own indignation at the crimes of Paris nor admiration for the virtues of Proteus.

There is another thread that runs through the story. Again and again, the stranger and hospitality are privileged. We read first of a Temple to Aphrodite as the stranger—a unique holy place which honors the outsider.[64] Another temple, this to Heracles, is the place of sanctuary to which Paris' slaves go, finding refuge and welcome.[65] When Proteus first characterizes Paris' crime, it is as a "crime against his friend" (i.e., against one who extended hospitality).[66] When it comes time to "sentence" Paris, Proteus refers to his practice of never putting a foreigner to death, in particular one forced by bad weather to seek refuge in Egypt. Proteus feels compelled to punish Paris "for the sake of your Greek *host*."[67] [Emphasis added.] The three days he gives Paris and friends to leave Egypt is apparently calculated to allow them to find anchorage elsewhere.[68] Putting all of this together it is hard to avoid the impression that the other, the stranger, is elevated above the native. Reciprocity and the equality it implied are fine for cross-cultural understanding, but when it comes to cross-cultural encounters and dealings, Herodotus privileges action founded on the presumption of hospitality. Side by side with a justice that cuts across boundaries and brings anyone, "no matter who," to task for his/her crimes, there is a concept of hospitality at play. Taken together justice and hospitality would seem to stand as two overarching principles that govern the encounters between natives and strangers.

Our present-day global scenario is riven by conflicts between cultures. Might the hospitality and justice exhibited in the Paris/Proteus story provide a pertinent Herodotan response? Readers must decide for themselves how pertinent these virtues are to our world. For its part, hospitality offers an alternative to a dynamic of distrust, suspicion, fear, and antagonism that might easily arise between native and foreigner. The hospitality of the host puts the guest-foreigner at ease. The alien is made to feel less alienated and more at home. The divide between the foreign and familiar becomes a threshold across which one may move. With some risk and much trust, the host opens his world to the stranger so that the stranger responds and reposes trust in the host.

In the Herodotan scheme, justice complements hospitality. One who violates the special trust that hospitality extends and builds—"no matter who . . ."—puts himself outside the pale of society—but, not beyond redemption: Paris is banished from Egypt but has time to seek refuge in another port. Custom does not dictate the judgment against the one who violates the bond of trust born of hospitality; nor does custom offer an excuse to the offender. Morality trumps mores. Justice is blind, except to what is right and wrong.

With some imagination one can see in the parable of Paris and Proteus the logic of confidence-building measures favored by those who practice conflict prevention and resolution today. Thus might an ancient scheme that addresses the one-on-one encounter of individuals have pertinence to a modern "face off" between opposing groups? If the story does not prove its contemporary social relevance, it certainly invites us to consider well its utility to our times.

In these concluding reflections on piety, morality, hospitality, and justice, we may have seemed to stray some distance from the domain of geography. It appears that we have shifted from geography to history, from description to prescription, and from the customs of particular cultures to universal ethical standards. Perhaps, in some sense, we have shifted attention. The boundaries these distinctions imply, however, need not confine our thinking and do not seem to have hampered Herodotan inquiry. For him, the earth is ever the setting and stage for the dramas that play out in history and engage one's sense of right and wrong. History "takes *place*." Just as the battles of Thermopylae and Salamis are shaped by the singular topographies to which the place names refer, so too, issues of hospitality and justice presuppose an encounter in time and *place* between native and stranger. Moreover, the difference that separates and distinguishes native and stranger is one of *place*. To be sure, place-bound mores have an immediately understood geographic sense. However, standards which transcend a particular time and place not only refer back to localities and boundaries which they exceed, but also point ahead to a horizon and openness which allows the local and bounded to emerge as such. Both terms, "boundary" and "horizon"—however abstractly we speak of them—unfailingly recall their geographical origins.

None of what was said earlier regarding the value of a situated earthbound understanding of the other is revoked by these latter considerations. Analogy and reciprocity remain indispensable tools for cross-cultural understanding. But Herodotan ethno-geography does not have the final word. The open-ended investigation of the human phenomenon in all its diversity across the globe—an admirable and useful project of its own—affords a propaedeutic, as Kant might say, to the ethical and historical issues that arise in the encounter and conflict of native and stranger. To his credit, Herodotus takes us beyond Herodotus. But if he addresses the ethical, this is not to say that he stipulates principles for right conduct or reflects explicitly on the good as philosophers after him would. He is yet Herodotus. His virtue is to see the human condition *in concreto*: peoples situated in their respective worlds, individuals formed and tested by their circumstances. For his earthbound, worldly understanding, virtue and crime, *hybris* and fate were palpable qualities of the fabric of human life. Perhaps it is this above all, his earthbound worldiness that makes for his appeal and enduring relevance.

Notes

1. Edith Hamilton, *The Greek Way* (New York: W. W. Norton & Co, 1983), 120.
2. Entry on Herodotus from *The Encyclopedia Britttanica, Eleventh Edition.* (Cambridge: Cambridge University Press, 1910-1911) from "Ancient History Sourcebook," http://wwwFord- aham.edu/halsall/ancient/eb1 1-herodotus.html, accessed July 23, 2002.
3. James Romm, *Herodotus* (New Haven, Conn.: Yale University Press, 1998), 50-51.
4. Herodotus, *The Histories,* trans. Aubrey de Selincourt (Baltimore: Penguin Books, 1954), 108.
5. Romm, *Herodotus,* 48.
6. Herodotus, *The Histories*, 106.
7. *Webster's Third New International Dictionary of the English Language* (Springfield, Mass.: G. & C. Merriam Co., 1966).
8. Herodotus, *Histories*, 105.
9. Herodotus, *Histories*, 299-307.
10. Richard Hartshorne, *The Nature of Geography: A Critical Survey of Current Thought in the Light of the Past, Annals of the Association of American Geographers,* Volume XXIX, Numbers 3 and 4 (Lancaster, Penn.: Association of American Geographers, 1939) 463.
11. G. W. F. Hegel, *Philosophy of Mind*, trans. William Wallace (Oxford: Clarendon Press, 1971), Sections 394-395, 40-51.
12. Hegel, *Mind*, 42.
13. Hegel, *Mind*, 41.
14. Hegel, *Mind*, 42-44.
15. Hegel, *Mind*, 43.

16. Hegel, *Mind*, 45.
17. Herodotus, *Histories*, 253-255.
18. Herodotus, *Histories*, 256.
19. Herodotus, *Histories*, 219.
20. Herodotus, *Histories*, 221.
21. Herodotus, *Histories*, 218.
22. Herodotus, *Histories*, 306.
23. Herodotus, *Histories*, 71.
24. Herodotus, *Histories*, 109.
25. Herodotus, *Histories*, 109.
26. Herodotus, Histories, 110.
27. Herodotus, *Histories*, 110-111.
28. Herodotus, *Histories*, 105.
29. Herodotus, *Histories*, 135.
30. Herodotus, *Histories*, 217.
31. Herodotus, *Histories*, 259.
32. Herodotus, *Histories*, 257.
33. Herodotus, *Histories*, 257-261.
34. Herodotus, *Histories*, 137.
35. Herodotus, *Histories*, 136.
36. Herodotus, *Histories*, 91.
37. Herodotus, *Histories*, 304.
38. *Webster's Third*.
39. Herodotus, *Histories*, 277.
40. Herodotus, *Histories*, 115-116.
41. Herodotus, *Histories*, 120.
42. Herodotus, *Histories*, 116.
43. Herodotus, *Histories*, 221.
44. Herodotus, *Histories*, 257.
45. Herodotus, *Histories*, 260.
46. Herodotus, *Histories*, 129. "The hippopotamus . . . has four legs, cloven hoofs like an ox, a snub nose, a horse's mane and tail, conspicuous tusks, a voice like a horse's neigh, and is about the size of a very large ox."
47. Herodotus, *Histories*, 122, 124.
48. Herodotus, *Histories*, 118-119.
49. Herodotus, *Histories*, 166.
50. Romm, *Herodotus*, 97.
51. Samuel P. Huntington, *The Clash of Civilizations and the Remaking of World Order* (New Delhi: Viking, Penguin Books India (P) Ltd., 1996), 19-21.
52. Herodotus, *Histories*, 127.
53. Herodotus, *Histories*, 71.
54. Herodotus, *Histories*, 108-109.
55. Herodotus, *Histories*, 31.
56. Herodotus, *Histories*, 10.
57. Friedrich Schliermacher, *On Religion, Speeches to its Cultured Despisers*, trans. Richard Crouter (Cambridge: Cambridge University Press, 1988), 104.
58. Herodotus, *Histories*, 57-65.
59. Herodotus, *Histories*, 518-519.
60. Herodotus, *Histories*, 143-147.
61. Herodotus, *Histories*, 144.

62. Herodotus, *Histories*, 146.
63. Herodotus, *Histories*, 147.
64. Herodotus, *Histories*, 143.
65. Herodotus, *Histories*, 143.
66. Herodotus, *Histories*, 144.
67. Herodotus, *Histories*, 144.
68. Herodotus, *Histories*, 144.

Chapter 2

Conceptualizing World Environmental History: The Contribution of Immanuel Wallerstein

Jason W. Moore[1]

> [F]actors of the physical environment . . . should be assessed and given their due weight. . . . [I]ntruding the variables of the physical environment does not undo our previous [social] analysis. It enriches it by adding a further element to help explain a historical conjuncture so consequential in the future history of the world.[2]

After some two decades of widespread concern over world ecology and sustainable development, world environmental history emerged as a small but significant field of inquiry in the 1990s.[3] A decade later, the enterprise remains weakly conceptualized. Its practitioners have been reluctant to engage in the theoretical labors that informed the conceptually-rich historical analyses pioneered by an earlier wave of world historians, primarily concerned with political economy rather than ecological crisis. Such under-theorization poses some substantial problems, since the enterprise of world history, even more so than local and regional history, poses difficult theoretical questions relating to, among other things, the nature of what constitutes a "world," the (ever-shifting) relations between "global" and "local" space, and the ways that environmental transformations are at once cause and consequence of large-scale socio-spatial transformations.

This paper tackles these dicey questions by turning to what environmental historians might regard as an unusual source, Immanuel Wallerstein's *The Modern World-System I: Capitalist Agriculture and the Origins of the European World-Economy in the Sixteenth Century*. Among environmentally-inclined world-systems analysts and globally-oriented environmental historians, the consensus holds that while *The Modern World-System* may be useful for matters of large-scale economic history, it is silent on the environment.[4] This consensus, however, betrays a casual reading of the text itself. A closer reading of *The Modern World-System*, I suggest, points to the volume's strong ecohistorical content, and its importance for conceptualizing a world environmental history that highlights capitalism's historical-geographical specificity.

This is a comradely critique, reconstructive rather than deconstructive. *The Modern World-System* deserves to be read not only for its limitations—the way it is usually read these days—but also for its profoundly suggestive treatment of agro-ecological transformations in the transition from feudalism to capitalism. However understated, *The Modern World-System*'s handling of socio-ecological dynamics in the crisis of feudalism and the transition to capitalism is unrivalled. While feudalism's agro-ecological contradictions were widely recognized prior to the mid-1970s,[5] Wallerstein's innovation was to link these contradictions to the emergence of capitalism, itself the agent of new and far-reaching ecological transformations.[6]

Immanuel Wallerstein, Environmental Historian?

The Modern World-System is remarkable in two respects—one sociological, the other geographical. In the first instance, it is a perceptive study of how world market formation shaped, and was shaped by, regional patterns of class conflict, state formation (and deformation), and associated modes of agricultural production—all of which were undergoing profound transformations precisely because the ecohistorical limits of the feudal system had been reached.

This world-historical *sociology* is complemented by a world-historical *geography*. Contrary to the stock criticism of the world-systems perspective, Wallerstein's handling of the local-global dialectic reveals a sophisticated tension between scales rather than a global determinism.[7] Indeed, *The Modern World-System* is very smart about space. Rather than begin with an abstract spatial entity (such as the nation-state, or even "Europe"),[8] for instance, Wallerstein emphasizes the production of space in ways that prefigured later Marxist theorizations of uneven development.[9]

Four moments of Wallerstein's historical geography of the transition to capitalism deserve special attention: equalization, expansion, divergence, and agro-ecological transformation. First, there was the process of *equalization* across space. Through the production of a new geographical scale—the capitalist world-economy—Europe's leading strata brought together formerly isolated or only loosely-articulated areas into a single division of labor. This new world-economy took shape as medieval Europe's multiple smaller world-economies merged, above all in the North Sea and the Mediterranean.[10] We can speak of equalization because the social relations encompassed by the new capitalist world-economy were increasingly (if unevenly) subordinated to the equalizing tendencies of the law of value, mediated through a competitive world market.

Second, there was the process of *expansion*. Geographical expansion, above all into the Americas, was essential to the resolution of feudal crisis in a way favorable to capitalist development. Even at this early date, the imperative of ceaseless geographical expansion emerges as the spatial corollary of ceaseless capital accumulation. The expansion of the scale of the world-economy was reinforced by the consolidation and expansion of the scale of the territorial states and their empires, which had suffered during the feudal crisis but gained renewed vigor through overseas expansion.

Third, as widely noted, there was the process of *divergence*. A new globalizing relation between core and periphery took shape, initially between eastern and western Europe and between western Europe and the Americas.

Finally, there was *agro-ecological transformation*. The rise of capitalism was part and parcel of a radical reshaping of world ecology, most dramatically in the new American and eastern European peripheries.

While the first three aspects of Wallerstein's historical geography are well known, the fourth is not. I propose to address these three moments of uneven development through this fourth optic, agro-ecological transformation. We will explore the ways that the emergence of a pan-European world-economy, stretching from the Baltic to the Americas, was at once cause and consequence of an epochal reorganization of world ecology. On the one hand, the emergence of capitalism as a response to feudal crisis by Europe's ruling strata was conditioned by the ecological contradictions of the feudal system (such as declining soil fertility). On the other hand, the likelihood of a successful transition to capitalism (rather than a reversion to feudalism or a world imperium), was itself predicated on the widening and deepening transformation of the earth so as to favor the generalization of commodity production.

Wallerstein's world-historical sociology and world-historical geography are dialectically bound by an ecohistorical interpretation of feudal crisis and capitalist transition. This interpretation is crystallized in his apt formulation of:

1) the crisis of feudalism as a "sociophysical" conjuncture[11] and
2) the emergence of capitalism as an epochal reorganization of "world ecology.[12]

I will address these two ecohistorical moments—one central to the crisis of feudalism, the other central to the emergence of a capitalist world-economy—in turn.

Linking the socio-ecological crisis of feudalism with capitalism's reorganization of world ecology is Europe's geographical expansion. For Wallerstein, feudalism's class contradictions in the crisis years of fourteenth and fifteenth centuries favored such expansion, which allowed medieval Europe's ruling strata to recoup its economic losses without engaging in a costly and perhaps fruitless struggle to reimpose serfdom on western Europe's insurgent peasantries. For geographical expansion to serve this purpose, it had to be profitable. The extension of the commodity production to frontier zones (silver, sugar), at first in the

Atlantic islands and then in the Americas, was the decisive moment of world ecological reorganization. Reading *The Modern World-System*, we learn that the processes of capitalist widening and deepening made possible, and *were made possible by*, radical and far-reaching environmental transformation, necessitating yet further geographical expansion in a ceaseless quest for uncommodified supplies of land and labor.

The Agro-Ecological Foundations of Feudal Crisis

Let us begin with the crisis of feudalism as "socio-physical" conjuncture. Wallerstein proposes that the "feudal system of social organization" could advance only so far before encountering insuperable limits. A system based on the political extraction of surplus offered few incentives for increased productivity. Because peasant surpluses were vulnerable to seizure by the seigniors, feudal property relations tended to limit the surplus available for investment in agricultural improvement. Economic expansion was therefore contingent upon geographical expansion. Seigniorial revenues increased as long as the population continued to grow, which meant that the amount of land under cultivation tended to expand, all other things being equal. And this was precisely the case between the eleventh and later thirteenth centuries.

Around 1300, however, feudalism appears to have overstepped the socio-ecological limits to continued expansion. The feudal organization of agriculture had begun to exhaust its land and labor power in the European heartland, and yield ratios (already quite low by world standards) stagnated or even declined. Meanwhile, settler expansion had brought more and more people onto less and less productive land at the geographical margins of the system. The margin of survival for European peasant agriculture was always razor thin. Relative overpopulation in the heartland, alongside overextension at the margins, undermined soil fertility and rendered fourteenth-century agriculture vulnerable to climate change. The agrarian recession of the later thirteenth century was already in motion prior to any epochal environmental shifts. But small shifts of any kind during periods of socio-ecological stress can produce big changes. In this sense, Wallerstein counts the arrival of chilly weather among the "cumulative woes" that "dealt a crushing blow to the already fragile demographic structure" of European feudalism.[13]

Also among feudalism's cumulative woes was epidemic disease. The decisive moment in the crisis of feudalism was the coming of the Black Death in 1348. The catastrophic impact of the Plague can be traced to "the chronic factor of resource strain involved in the feudal system of social organization."[14] Late feudalism's eco-geographical overextension and overexploitation not only rendered agriculture highly vulnerable to climate change, but also biologically impoverished the peasantry. Malnutrition rendered the mass of the population vulnerable to disease. In a system where the social reproduction of the ruling class hinged upon a growing population in order to sustain seigniorial revenues, the Black

Death quickly transformed the agrarian depression of the early fourteenth century into a terminal crisis of the feudal system.

At base, the crisis of feudalism was one of declining seigniorial revenues, owing to the demographic decline and the resulting enhanced bargaining position of the peasantry. The crisis of the seigniors led in short order to the crises of the political institutions of Europe, especially the states and the Church. This was also a moment of crisis for capitalists based in the city-states, who faced declining returns on trade and manufacturing, as the economic base of European civilization crumbled. The conjuncture of these multiple crises would play a key role in the resolution of feudal crisis—a crisis which, as we know, resulted in the transition to capitalism rather than to another tributary system.

Wallerstein's analysis of the transition pivots on the relationship between class structure and the labor-land ratio. Where population density and urbanization remained *relatively* high, as in western Europe, the peasantry's power was augmented proportionately. The new balance of class forces in western Europe therefore favored the rise of the yeoman farmer and intensive agriculture. Here the peasantry's class power precluded the reimposition of serfdom, which in the short run exacerbated the ongoing crises of western Europe's ruling strata. Where population density and urbanization was relatively low, as in eastern Europe and the Americas, extensive agriculture developed on the basis of "coerced cash-crop labor" (serfdom) or outright slavery. This difference was the result at once of the differential possibilities for "*effective* resistance," owing to varying degrees of urbanization and population densities, and of the differential commercial opportunities presented by the relative availability of land: "If there is plenty of land, one can make do with relatively inefficient means of production. One can engage in extensive agriculture. One can use slaves or coerced cash-crop laborers [serfs]."[15]

The intensification of western European agriculture and the spread of extensive, cash-crop agriculture in eastern Europe and the Americas were complementary processes. The spread of coercive modes of labor control in the new peripheries—especially slavery in the Atlantic and the "second serfdom" in eastern Europe—was possible to the extent that a significantly larger world market for cereals and sugar emerged in the fifteenth and sixteenth centuries. But this larger market could develop only on the basis of industrialization and agricultural innovation in the emergent core regions. High value-added industrial and agricultural sectors, in the Low Countries for instance, were sustainable only insofar as reliable grain supplies could be obtained for the cities.[16] Moreover, industrial expansion required horsepower, which meant that arable land had to be converted to pasturage, which meant that workers had to be fed with (even more) imported grain. Rising core demand for grain, in turn, sent prices and profits upwards, thereby locking the peripheral Baltic into an expanding world-scale division of labor.[17] "Hence, the process of agricultural innovation fed rather than foreclosed the necessity of expansion."[18]

The Centrality of Geographical Expansion

Geographical expansion was not only necessary—that is, necessary if the feudal crisis was to be resolved in favor of capitalist transition—but practical. It was necessary because the possibilities for inner expansion were limited—not by population but by social structure; outer expansion in contrast was practical because of the proximity of the Atlantic islands and the Americas and feasibility of cash-crop production in these new areas.[19]

The central problem was not too little land but too much:

> There was physical room for the population, even the growing population. *Indeed that was part of the very problem that led to expansion*. The physical room was one element in the strength of the peasantry vis-à-vis the nobility, and hence one factor in the decline of seigniorial revenues, in the crisis of feudalism. . . . What the nobility (and the bourgeoisie) needed . . . was a more tractable labor force. The size of the population was not the issue; it was the social relations that governed the interaction between the upper and lower classes. . . . Europe needed a larger land base to support the expansion of its economy, one which could compensate for the critical decline in seigniorial revenues and which could cut short the nascent and potentially very violent class war which the crisis of feudalism implied.[20]

Transatlantic expansion was the path of least resistance, given the reality of overlapping crises, which pushed together interests which had hitherto been at odds. Because it had empowered the peasantry, the feudal crisis pushed together the interests of Europe's ruling strata—the states, the seigniors, and the city-state capitalists—to favor outer rather than inner expansion. "The only solution that would extract western Europe from decimation and stagnation would be one that would expand the economic pie to be shared, a solution which required, given the technology of the time, an expansion of the land area and population base to exploit."[21] Wallerstein shows that the creation of a capitalist world-economy was the outcome of a conjuncture of crises through which the interests of these three major groups—the territorial states, the seigniors, and the city-state capitalists—converged to favor overseas expansion.

First, the territorial states, which had made great strides between the eleventh and fourteenth centuries—owing to increased revenues from "internal" expansion and the politico-military unification which resulted from the Crusades—now suffered greatly from the economic contraction, which began even before the Black Death. Beginning in the fourteenth century, the territorial states faced a deepening "liquidity crisis" as they struggled to exact higher taxes from the peasants in the interests of waging war.[22]

Second, the seigniors faced a deepening crisis in the wake of the Black Death. The downward readjustment of labor-land ratios effected several crucial changes in the balance of social forces, particularly in western Europe. First, the economic contraction in the countryside, which began in the later *thirteenth* century, produced rising social unrest. Peasant revolts grew more frequent, and extended their theater of operations from villages to larger regions; urban unrest increased as well.[23] Second, once population contraction set in, at first slowly through the increasing frequency of famine and then rapidly with the coming of the Black Death, revenues declined. Third, declining population led to increased competition among the seigniors for peasant labor.

In this situation, an internal fix of the seigniorial crisis was hardly a possibility, given the widespread abandonment of villages and cultivated land that followed in the wake of the Black Death. Whatever inner expansion occurred was very largely an expansion of pasturage rather than arable land, a move that militated against a rapid population recovery.[24] Sheep farming especially not only required fewer hands relative to agriculture, it yoked the seigniors to the world market, who were as a consequence inclined to support measures that favored the further expansion of that market. The resulting widespread displacement of cereal agriculture by animal husbandry not only entailed a deepening of the world-economy's division of labor, but biased it in favor of further expansion. It was no mere coincidence that Castile, itself one of the two great sheep farming areas of Europe, led the conquest of the New World, and established sheep farming almost immediately upon arrival. "[S]heep ate men, in middle America just as in England."[25] And it was not only Europeans who were "eaten" by sheep and other livestock. Indian cultivation directly suffered from trampling livestock, an important contributing factor to their great demographic collapse of the sixteenth century.[26]

Meanwhile, owing to increasing seigniorial competition, declining revenues, and relatively high urbanization after the Black Death, western Europe's peasantry waged the class struggle much more effectively than heretofore, squeezing the seigniors, who in turn squeezed the states, who were forced to recognize the former's voice in policy-making. Ultimately, the seigniors could expand their revenues only so far as "their" states prospered, and during the crisis of feudalism such prosperity was limited to the extent that inner expansion was privileged over outer expansion—precisely because of the peasantry's strength. Thus an uneasy compromise prevailed which was ultimately resolved in favor of territorial state power and overseas expansion.

Third, the city-states also experienced particular and general contradictions that favored geographical expansion. Presumably, the city-state capitalists faced a contracting market once seigniorial revenues began to decline. Pushed out of the eastern Mediterranean by Venice, Genoa provided the capital necessary for Iberian expansion.[27] And it didn't hurt that geographical expansion "minimized . . . [the] potential for internal disorder . . . [posed by the] urban semiproletariat."[28]

Agro-Ecological Transformations in the Transition to Capitalism

The convergence of interests in favor of overseas expansion favored the transition to capitalism. In the abstract, however, expansion tells us little about the capitalist reorganization of world ecology. European expansion was also a process of consolidation, which brought together the existing North Sea and Mediterranean world-economies into a relatively unified division of labor.[29] Even without an immediate increase in agricultural productivity, the enlarged scale of the new world-economy provides its ruling strata with a greatly enlarged resource base. This integration of Europe's world-economies was complemented by geographical expansion and core-periphery polarization. Not only the economic surplus, but the *ecological surplus*, that derived from Europe's expanding division of labor was "unequally consumed."[30] The European capitalism's resource base was not only *absolutely* larger than its predecessor world-economies. Thanks to the inequalities and incentives built-in to capitalist development, it was *relatively* larger—much larger.

Among these resources was American bullion. Among the more intriguing aspects of Wallerstein's interpretation of American bullion's importance to the emergent world-economy is its ecological dimension. Why was bullion so important to the "vast but weak" world-system? "It sustained the thrust of the expansion, *protecting this weak system against the assaults of nature.*"[31] Gold and silver protected the original accumulation of capital against the devaluing impact of poor harvests. Would Braudel's "Age of the Genoese"[32] have been possible without bullion to lubricate the trade between northern and southern Europe, and to insure Genoa against the threat of famine?[33]

Equally important in "protecting this weak system against the assaults of nature" were New World crops. By itself, Europe's biodiversity was probably insufficient to sustain significant population growth. "Without the American crops, Europe might not have been able to carry such heavy populations as she later did, and the Old World tropics would not have been so quickly developed."[34]

For Wallerstein, "the thrust of [European] expansion" arose from the ascendant logic of capital accumulation under conditions of resource strain—conditions largely created by capital's ascent and feudalism's dissolution. "What western Europe needed in the fourteenth and fifteenth centuries was food (more calories and a better distribution of food values) and fuel."[35] The drive for fuel and food—especially wood, wheat, and sugar—reinforced capital's tendency towards uneven development, in the case of western and eastern Europe transforming the latter's small differences into large and durable inequality,[36] and creating new peripheries in the Atlantic islands and the Americas.

This development of an unequal world division of labor, in turn, created new capitalist efficiencies of specialization between arable and pasture, and between agriculture and industry. Predictably, these processes encouraged capitalist monoculture and the prioritization of short-run profits over sustainability. In England, "the orientation of these town bourgeois was toward short-run profit . . . which had the effect of desolating the land over the following [sixteenth] century."[37]

In Spain, a similar process was ongoing. Triennial rotations were forsaken, leading to soil exhaustion. "This situation led, in the years from 1570 to 1630, to the search for new domains, which were then similarly exhausted."[38] Wallerstein's analysis gets more interesting from this point, however. It turns out that not only did primitive accumulation lead to ecological degradation, but that in Spain, ecological degradation led to a new phase of primitive accumulation:

> [The] "aridification" of the land hit particularly the small producers and led to further land concentration. While this process of exhausting the land led to inability to export, land concentration that resulted from it led to a further *monetarization* of productive relations, since "the inability of peasants . . . and villagers to provide for their own subsistence by working their own land enlarged the internal market."[39] [Emphasis added.]

Next to food, forest products were "the other great basic need" in the early modern European world-economy. Alongside sugar, they were the era's "continuing 'growth' crop."[40] England's invasion and subsequent colonization of Ireland after 1600 was ecologically as well as socially devastating. Ireland's "woods were used up to supply England with timber," as well as to deprive resisters of forest cover from English cavalry.[41] Whereas one-eighth of the Emerald Isle was forested in 1600, the forest had "virtually disappeared by 1700."[42] In Spain, the forests receded under the pressure of a rising sheep population that was a direct response to growing world market demand.[43]

In the sixteenth-century Mediterranean, the "wood famine" was so great that "Nordic wood arrived in Seville with boats filled to the brim with planks and beams."[44] Growing demand led inexorably to the "slow but steady deforestation of western Europe, Italy, and Spain, as well as Mediterranean islands. Oak became especially scarce."[45] In France, the iron industry's devastation of forests sparked peasant revolts by the 1730s.[46] Dwindling timber supplies in the core and semiperiphery of the early modern world-economy led to the expansion of what we might call "cash-crop" forestry in the Baltic region, which by "the 16th century . . . had begun to export wood in large quantities to Holland, England, and the Iberian peninsula."[47] As a consequence, by the time of the Thirty Years' War (1618-48)—itself a major source of "ecological stress"[48]—the "reckless exploitation" of Poland's forests had "produced a desert in the woods."[49]

Perhaps most important is the case of sugar. "Unlike wheat, cattle, and silver, sugar was not involved in the problem of oversupply" in the early modern world-economy.[50] Wallerstein links together: (1) the production of peripheral space and degraded landscapes through the plantation system,[51] (2) the degradation of the soil and the expansion of the world-economy and, (3) the degradation of the soil and the degradation of the worker. In the first case, Wallerstein argues that the trend towards monoculture was most evident in the periphery. Plantation-driven monocultures were the most dramatic, and quite possibly the most important, way in which peripheral landscapes were produced in the emergent capitalist world-economy. "The trend in the *core* was towards variety and specialization [in high value-added crops], while the trend in the periphery was toward monoculture."[52]

Secondly, sugar's "eternal problem was ecological exhaustion and the necessity to find virgin zones to exploit."[53] In a passage that seriously undermines the critique of circulationism directed against Wallerstein, he explicitly rejects Pierre Chaunu's economistic explanation[54] of Brazil's early seventeenth-century sugar revolution in favor of an ecological one:

> Is it not easier to explain this expansion by the previously discussed relationship between the fairly rapid rate of ecological exhaustion and world demand and conclude that as a product of the world-economy, sugar was less subject to secular swings than were wheat and silver?[55]

Thirdly, Wallerstein sees a dialectical connection between the degradation of the soil and the degradation of the worker: "The [largely sugar] monocultures imposed on the Mediterranean and Atlantic islands ravaged them, pedologically and in terms of human population. Their soils were despoiled, their populations died out."[56] With sugar, we have a "very lucrative and demanding product, pushing out wheat but then exhausting the soil, so that it required ever new lands (not to speak of the manpower exhausted by its cultivators)."[57] In *The Modern World-System II*, Wallerstein extends this line of analysis to eastern Europe's agricultural crisis in the early seventeenth century: "Production was increased 'by deviating from the fundamental principles of rotation in tilling the soil,' which over time exhausted the soil. Exhausting the *men and the soil* maintained a level of total production for 50-60 years, but it was a self-containing [and self-defeating!] method."[58]

In both cases (sugar and wheat), Wallerstein links capitalism's degradation of the soil with the degradation of the worker. For Wallerstein, as for Marx,[59] the transformation of nature *is* a labor process. The degradation of nature *is* the degradation of the worker, and occurs *only through* the degradation of the worker under the law of value.[60]

For now, it bears emphasizing that this degradation of the worker could occur both at the point of production and through the instrument of world trade—through relations of production and relations of consumption (socio-biological reproduction).[61] Hence, the "ecologically driven" westward movement of sugar, first to the Atlantic islands and thence to the Americas, led to the emergence of a new capitalist and racialized slave regime based on African laborers. But why Africans as the new slaves? Because in Africa, capitalists found a nearby area from which it could extract slaves without concern for its negative economic impact on the region. Moreover, capitalist enterprise in the Americas was enormously destructive of labor: "sheer exhaustion of manpower, especially in the mines, must . . . have been significant."[62] The same could be said of sugar cultivation—the relatively low cost of slave labor was perfectly sutied for sugar, whose (generally) low skill requirements were matched by equally high mortality rates, high even by the standards of New World slavery.

It was not only slaves who suffered bodily in the transition to capitalism, however. In terms of Europeans' food consumption, the shift from arable to pasturage at the beginning of the long sixteenth century (1450-1640) assumes added importance. The Black Death reduced labor-land ratios, and thereby allowed for more land per capita. This should have allowed for a fairly rapid demographic recovery, all other things being equal. But equal they were not. "[A]lthough fewer men should have meant more food since the landmass remained the same, it also meant a shift to pasturage and hence a reduction of caloric output."[63] This reduction was of course unequally distributed. "European workers paid part of the costs of European economic development" through a deteriorating diet.[64] Europe experienced a "partial 'decerealization'. . . in favour of animal husbandry . . . [thereby] altering the pattern of consumption," which among other things meant that the bulk of the population suffered "scarcely less than previous centuries from death and famine"[65]:

> The most important thing to note about pasturage in the 16th century, especially livestock, was that it was becoming increasingly a regionally specialized activity. *More cattle here, an advantage to large landowners, also meant less cattle elsewhere, which often meant a reduction in peasant consumption of meat and dairy producers, a deterioration in the diet.*[66] [Emphasis added.]

A new dietary regime was therefore one means by which bodily space was articulated with the broader town-country division of labor in the transition to capitalism.

The Modern World-System: Implications for World Environmental History

The reading of *The Modern World-System* proposed here suggests a renovated world-historical perspective capable of illuminating the ways in which world market formation, the geographical expansion of capital and commodity production, and agro-ecological transformation interacted at *multiple geographical scales* (Wallerstein's "layers within layers")—at the scale of the world-economy, yes, but equally as a force behind uneven regional development and socio-ecological change *at the point of production* (including human bodies themselves). From this perspective, we can account for capitalism's oscillation between "fixity and motion" in a way that goes beyond an analysis of the circulation of monetary capital and its relation to the production of built environments—historical geographical relations that at first liberate and then imprison accumulation.[67] Indeed, we might identify among the sources of the system's cyclical alternations of de- and re-territorialization—between the "space of flows" and the "space of places"—as expressions of capital's relation with nature.

Wallerstein's contribution to the environmental history of the modern world, then, begins but does not end with the notion that environmental transformations occur within those very large spaces called world-systems. Beyond Wallerstein's quiet recognition that the appropriation of metabolic flows has always been central to capital accumulation, and that these flows respect no "national" boundaries, we can identify in *The Modern World-System* two significant contributions to world environmental history. First, in showing that the consolidation of a capitalist world-economy was contingent upon a far-reaching reorganization of world ecology, Wallerstein makes a powerful argument that the chief cause of early modern environmental transformation was the accumulation of capital and the generalization of commodity production. Long before the advent of fossil fuel technology and the steam engine, and during an era when population density declined precipitously in the Atlantic world,[68] capitalists were engaged in simplifying and degrading laborers and landscapes alike. As a result, capital was forced to look perpetually for new sources of land and labor, effectively globalizing environmental problems in way that sets early European capitalism apart from civilizations in Afro-Eurasia and the Americas. Wallerstein demonstrates that *capital* at once precedes and precipitates the waves of industrialization and population growth that are so often identified as the chief culprits of global ecological crisis today.[69]

Second, Wallerstein prefigures both Donald Worster's arguments for the centrality of agriculture in ecohistorical analysis and J. Donald Hughes, call for explanations of "the background of the world market economy and its effects on the global environment."[70] Taking the transition from feudalism to capitalism as its overarching problematic, *The Modern World-System* deploys an essentially agrarian geographical conception of capitalism among whose many virtues is an emphasis on the relation between commodity production, environmental transformation, and an emergent capitalist world market. Like Robert Brenner,[71] and in contrast to Fernand Braudel and Giovanni Arrighi,[72] Wallerstein's analytical strategy prioritizes *agrarian* capitalism over its urban forms. The emphasis on agrarian capitalism is for the most part right on target, illuminating decisive economic sectors and ecological transformations in early modern capitalism. Unless and until agriculture could be drawn into the orbit of capital accumulation and subjected to the competitive pressures of the law of value, capitalism could not develop, much less get started.[73]

The Modern World-System, then, suggests a powerful, ecohistorical restatement of traditional versions of the transition from feudalism to capitalism. The emphasis on commodity production is especially important. While feudalism degraded the environment in significant ways, such transformations did not lead to the globalization of environmental problems within the feudal system of production. Wallerstein helps us see that underlying the fundamental ecohistorical difference between capitalism and feudalism is the role of commodity production in the two systems. To be sure, there was commodity production under feudalism. There were, for example, important antecedents of the modern plantation system in the medieval Mediterranean. But however *widespread* this commodity production may have been, there was no ineluctable tendency towards its *generalization*. Why? Because a society organized around the progressive generalization of commodity production undermines relations of domination based on tribute. Social strata that benefit from this system are likely to oppose any change that might favor generalized commodity production. In the end, however, they had no choice. The crisis of feudalism led to a convergence of interests among Europe's ruling strata in favor of a significant expansion of commodity production, most dramatically in the New World.

Geographical expansion was so important because it incorporated vast new agrarian zones—with weak or non-existent peasantries—into the emergent pan-European world-economy. Creating new agrarian spaces for commodity production, outside the peasantry's western European stronghold, encouraged Europe's beleaguered ruling strata to reshape the town-country division of labor in ways that favored capitalist development. Above all through the agency of the sugar plantation and the massive silver mining enterprises of Potosi and central Mexico, early capitalism spearheaded the generalization of commodity production through overseas expansion, the fruits of which fueled capital accumulation in Europe's leading cities.[74] In turn, this accumulation made possible the extension of capitalist town-country relations and the deepening of commodity relations in the European countryside. Control over American silver during the era of the Price Revolution, for instance, allowed Dutch merchants to subordinate Poland's grain producing landlords to the world market, essentially creating a system of "international debt peonage."[75]

Mining enterprises and sugar plantations both funneled ecological as well as monetary wealth from the increasingly-distant rural areas into the cities. Thus did the "development of underdevelopment" and capitalism's "*metabolic rift*" form a dialectical unity. Perhaps nowhere was this more evident than in the emergent sugar plantation complex. First in the Atlantic islands and later in Brazil, the capitalist sugar plantation in the fifteenth and sixteenth centuries ravaged both the island landscapes it occupied and the laborers who worked them. With sugar, we have a classic instance of capitalism's metabolic rift—whereby the products of the countryside (especially but not only in the periphery) flow into the cities, which are under no obligation to return the "waste" products to the point of production.[76] Nutrient cycling is compromised, as nutrients are pumped out of rural areas and transferred to urban centers. From this standpoint, the expansion of capitalist agriculture rests upon the "increasing overdraft on the fertility of our soils."[77] As a consequence, the wave of European expansion inaugurated by the crisis of feudalism set in motion new ecological crisis tendencies on a world scale. With the creation of a world market and a trans-Atlantic division of labor in the sixteenth century, feudalism's localized ecological problems gave way to capitalism's globalization of environmental problems. With the transition to capitalism, the nutrient cycling of local ecosystems was radically disrupted as these were incorporated into a new division of labor, and localized relations gave way to progressively globalized relations between the country and the city.[78]

The Modern World-System serves as a powerful reminder of how accumulation is always embedded in, and transformative of, the natural environment—an environment that includes human beings as well as landscapes. This is a broad conception of environmental history, which recognizes that landscapes are transformed by human agency, and that the agents of this transformation are variously liberated and degraded, that provides some means of distinguishing the historical specificity of capitalism's (and feudalism's) environmental transformations—a distinction that typically eludes environmental historians.[79] Wallerstein's careful attention to the production of space; the connections between environmental degradation, social crisis, and spatial expansion; the links among the world-scale accumulation of capital, the degradation of the soil, and the destruction of labor-power—these and more provide valuable insights into the ways that historical capitalism at once produces, and is produced by, the physical environment.

Notes

1. Address correspondence to: Jason W. Moore, Department of Geography, University of California, Berkeley, CA 94720; jasonwmoore@earthlink.net or jasonwmoore @hotmail.com. This paper would not have been possible without the support and encouragement of Diana Carol Moore Gildea, who, along with Brett Clark, read and commented upon successive drafts of this paper. I would also like to thank Giovanni Arrighi, Ben Brewer, John Bellamy Foster, Peter Grimes, David Harvey, Ho-fung Hung, Yukio Kawano, Jordan Leonard, and Nari Rhee for useful discussions. This draft February 25, 2003.

2. Immanuel Wallerstein, *The Modern World-System I* (New York: Academic Press, 1974), 33, 36.

3. Prefiguring the world environmental history boom of the 1990s were, among others, Lester J. Bilsky, ed., *Historical Ecology: Essays on Environment and Social Change* (Port Washington, N.Y.: Kennikat Press, 1980); Piers Blaikie and Harold Brookfield, *Land Degradation and Society* (London: Methuen, 1987); Fernand Braudel, *The Mediterranean and the Mediterranean World in the Age of Philip II* (2 vols., New York: Harper and Row, 1972-1973); Braudel, *The Structures of Everyday Life* (New York: Harper and Row, 1981); L. Brockway, *Science and Colonial Expansion: The Role of the British Botanic Gardens* (New York: Academic Press, 1979); Alfred W. Crosby, *The Columbian Exchange: Biological and Cultural Consequences of 1492* (Westport, Conn.: Greenwood Academic Press, 1972); Crosby, *Ecological Imperialism* (Cambridge: Cambridge University Press, 1986); Frank Perlin, *A Forest Journey: The Role of Wood in the Development of Civilization* (Cambridge, Mass.: Harvard University Press, 1989); John F. Richards and Richard P. Tucker, eds., *World Deforestation in the Twentieth Century* (Durham, N.C.: Duke University Press, 1988); Carl Sauer, *Selected Essays, 1963-1975* (Berkeley, Calif.: Turtle Island Foundation, 1981); William L. Thomas, Jr., ed., *Man's Role in Changing the Face*

of the Earth (Chicago: University of Chicago Press, 1956); Richard P. Tucker and John F. Richards, eds., *Global Deforestation in the Nineteenth Century World Economy* (Durham, N.C.: University of North Carolina Press, 1983); Jack Westoby, *Introduction to World Forestry* (Oxford: Basil Blackwell, 1989); Donald Worster, "World without Borders: The Internationalizing of Environmental History," in *Environmental History*, ed. Kendall E. Bailes (Lanham, Md.: University Press of America, 1985), 661-669; Worster, ed., *The Ends of the Earth: Perspectives on Modern Environmental History* (Cambridge: Cambridge University Press, 1988).

A representative sampling of modern world environmental history (ca. 1400 to the present) over the past decade includes Stephen Boyden, *Biohistory: The interplay between human society and the biosphere* (Paris: UNESCO, 1992); Ciccantell, Paul, and Stephen Bunker. (1999) "Economic Ascent and the Global Environment: World-Systems Theory and the New Historical Materialism," in *Ecology and the World-System*, eds. Walter L. Goldfrank, D. Goodman, and A. Szasz (Westport, Conn.: Greenwood, 1999), 107-122; Sing Chew, *World Ecological Degradation* (Walnut Creek, Calif.: Alta Mira Press, 2001); Alfred W. Crosby, *Germs Seeds, and Animals: Studies in Ecological History* (Armonk, N.Y.: M. E. Sharpe, 1994); Mike Davis, *Late Victorian Holocausts* (London: Verso, 2001); Jared Diamond, *Guns, Germs, and Steel: The Fates of Human Societies* (New York: Norton, 1997); Richard Drayton, *Nature's Government: Science, Imperial Britain, and the 'Improvement' of the World* (New Haven, Conn.: Yale University Press, 2000); Wilma A. Dunaway, "Incorporation of mountain ecosystems into the capitalist world-system," in *Review* 19 (4, 1996): 355-381; Mark Elvin and Liu Ts'ui-jung, eds., *Sediments of Time: Environment and Society in Chinese History* (Cambridge: Cambridge University Press, 1998); John Bellamy Foster, *The Vulnerable Planet* (New York: Monthly Review Press, 1994); Harriet Friedmann, "What on Earth is the Modern World-System? Foodgetting and Territory in the Modern Era and Beyond," in *Journal of World-Systems Research* 6 (2, Sum., 2000): 480-515; Tom Griffiths and Libby Robin, ed., *Ecology and Empire: Environmental History of Settler Societies* (Seattle: University of Washington Press, 1997); Richard H. Grove, *Green Imperialism: Colonial Expansion, Tropical Island Edens and the Origins of Environmentalism* (Cambridge: Cambridge University Press, 1995); Richard H. Grove, *Ecology, Climate, and Empire: Colonialism and Global Environmental History, 1400-1940*. (Cambridge: White Horse Press, 1997); Richard H. Grove and John Chappell, eds., *El Nino, History, and Crisis: Studies from the Asia-Pacific Region* (Cambridge: White Horse Press, 2000); Richard H. Grove, Vinita Damodaran, and Satpal Sangwan, eds., *Nature and the Orient: The Environmental History of South and Southeast Asia* (New Delhi: Oxford University Press, 1998); Yrjo Haila and Richard Levins, *Humanity and Nature: Ecology, Science, and Society* (London: Pluto, 1992); J. Donald Hughes, ed., *The Face of the Earth: Environment and World History* (Armonk, N.Y.: M.E. Sharpe, 2000); Calestous Juma, *The Gene Hunters: Biotechnology and the Scramble for Seeds* (Princeton, N.J.: Princeton University Press, 1989); Kozo Mayumi, "Temporary Emancipation from the Land: From the Industrial Revolution to the Present," in *Ecological Economics* 4 (1, 1991): 35-56; J. R. McNeill, *Something New Under the Sun: An Environmental History of the Twentieth Century* (New York: Norton, 2000); Clive Ponting, *A Green History of the World* (New York: St. Martin's Press, 1991); Vaclav Smil, *Energies: An Illustrated Guide to the Biosphere and Civilization* (Cambridge, Mass.: MIT Press, 1999); B. L. Turner II, William C. Clark, Robert W. Kates, John F. Richards, Jessica T. Mathews, and William B. Meyer, eds., *The Earth as Transformed by Human Action: Global and Regional Changes in the Biosphere over the Past 300 Years* (Cambridge: Cambridge University Press, 1990); Helen Wheatley, ed., *Agriculture, Resources Exploitation, and Environmental Change* (Brookfield,

Vt.: Variorum/Ashgate, 1997). A useful if brief review essay is offered by J. Donald Hughes, "Global Dimensions of Environmental History," in *Pacific Historical Review* 70 (1, 2001): 91-101.

4. Stephen G. Bunker, *Underdeveloping the Amazon* (Urbana: University of Illinois Press, 1985); Chew, *World Ecological Degradation*; Friedmann, "What on Earth is the Modern World-System?"; Groves, *Green Imperialism*; J. Timmons Roberts and Peter Grimes, "Extending the world-system to the whole system: Toward a political economy of the biosphere," in *Ecology and the World-System*, eds. Walter L. Goldfrank, D. Goodman, and A. Szasz (Westport, Conn.: Greenwood Press, 1999), 59-83; David A. Smith, "Uneven Development and the Environment: Toward a World-System Perspective," in *Humboldt Journal of Social Relations* 20 (1, 1994): 151-75; Worster, "World Without Borders."

5. Perry Anderson, *Passages from Antiquity to Feudalism* (London: New Left Books, 1974); Harry Miskimin, *The Economy of Early Renaissance Europe, 1300-1460* (New York: Prentice-Hall, 1969); M. M. Postan, *Medieval Economy and Society*, (New York: Pelican, 1972).

6. For further discussion, see Jason W. Moore, "Environmental Crises and the Metabolic Rift in World-Historical Perspective," in *Organization & Environment* 13 (2, 2000): 123-158; "Sugar and the Expansion of the Early Modern World-Economy," in *Review* (Fernand Braudel Center) 23 (3, 2000); "The Crisis of Feudalism: An Environmental History," in *Organization & Environment* 15 (3, 2002); "Nature and the Transition from Feudalism to Capitalism," in *Review* 26 (2, 2003); "*The Modern World-System* as Environmental History? Ecology and the Rise of Capitalism," *Theory and Society*, forthcoming.

7. For one of the more widely cited criticisms along these lines, see Steve J. Stern, "Feudalism, Capitalism, and the World-System in the Perspective of Latin America and the Caribbean," in *American Historical Review* (93, 1988): 829-873.

8. "But 'Europe' must not be reified. There was no central agency which acted in terms of these long-term objectives. The real decisions were taken by groups of men acting in terms of their immediate interests" (Wallerstein, *The Modern World-System I*, 51).

9. See David Harvey, *Spaces of Hope* (Berkeley: University of California Press, 2000); Neil Smith, *Uneven Development* (Oxford: Blackwell, 1984); Edward W. Soja, *Postmodern Geographies* (London: Verso, 1989).

10. Wallerstein, *The Modern World-System I*, 68, 77.

11. Wallerstein, *The Modern World-System I*, 35.

12. Wallerstein, *The Modern World-System I*, 44.

13. Wallerstein, *The Modern World-System I*, 34.

14. Wallerstein, *The Modern World-System I*, 35.

15. Wallerstein, *The Modern World-System I*, 104, 100-101, 112.

16. "That one or two percent [of Europe's total consumption provided by Baltic grain] was nevertheless exceptionally important, both because of the prosperity it brought to such seafarers as the Dutch and because it represented the margin of survival for capital cities like Lisbon" (Charles Tilly, "Food Supply in Early Modern Europe," in *The Formation of National States in Early Modern Europe*, ed. Charles Tilly (Princeton, N.J.: Princeton University Press, 1975), 416 quoted in Wallerstein, *The Modern World-System I*, 76.

17. Wallerstein, *The Modern World-System I*, 75-76.

18. Wallerstein, *The Modern World-System I*, 42-43.

19. I use the terms "inner" and "outer" expansion, following Terence K. Hopkins and Immanuel Wallerstein, "Patterns of development of the modern world-system," in *Review*

1 (2, 1977): 125: "The literature on agricultural history has indicated a clear pattern over time of 'inner' expansion, in the sense that not all the areas physically located inside the outer boundaries of the world-economy had necessarily been from the outset involved in the social economy. There were 'subsistence redoubts.' It is clear that, as a process, the incorporation of areas at the outer edges and the areas that were redoubts inside it were the same phenomenon economically, even if it had a different definition juridically and perhaps different prerequisites politically."

20. Wallerstein, *The Modern World-System I*, 48, 51, emphasis added.

21. Wallerstein, *The Modern World-System I*, 24.

22. Wallerstein, *The Modern World-System I*, 21.

23. Wallerstein, *The Modern World-System I*, 24.

24. Wallerstein, *The Modern World-System I*, 22, 35-36.

25. Wallerstein, *The Modern World-System I*, 188, 36, n. 78.

26. Wallerstein, *The Modern World-System I*, 89-90.

27. Wallerstein, *The Modern World-System I*, 49-50.

28. Wallerstein, *The Modern World-System I*, 52.

29. Wallerstein, *The Modern World-System I*, 68, 77.

30. Wallerstein, *The Modern World-System I*, 42.

31. Wallerstein, *The Modern World-System I*, 76, emphasis added.

32. Braudel, *Perspective of the World*.

33. See Wallerstein's approving quotation of Gustaf Utterstrom, "Climate Fluctuations and Population Problems in Early Modern Europe," *Scandinavian Economic History Review* 3 (1, 1955), 44: "The food situation in the Mediterranean area would in all probability have been much more serious if the flow of precious metals from America had not provided means of payment for the large purchases of grain" (quoted in *The Modern World-System I*, 217, n. 252).

34. G. B. Masefield, "Crops and Livestock," in *The Cambridge Economic History of Europe IV*, eds. E.E. Rich and C.H. Wilson (Cambridge: Cambridge University Press, 1967), 276, quoted in Wallerstein, *The Modern World-System I*, 44-45, n. 108.

35. Wallerstein, *The Modern World-System I*, 42.

36. Wallerstein, *The Modern World-System I*, 98-99, 111-12, 121-22, 129.

37. Wallerstein, *The Modern World-System I*, 107.

38. Wallerstein, *The Modern World-System II* (New York: Academic Press, 1980), 146.

39. Wallerstein, *The Modern World-System II*, 146.

40. Wallerstein, *The Modern World-System I*, 44-45; *The Modern World-System II*, 161-62.

41. Wallerstein, *The Modern World-System I*, 281, 281, n. 276.

42. Wallerstein, *The Modern World-System I*, 281.

43. Wallerstein, *The Modern World-System I*, 193, n. 138.

44. Fernand Braudel, *Le Mediterranee et le monde Mediterraneen à l'epoque de Philippe II*, 2e ed. revue et augmente. 2 vol. (Paris: Lib. Armand Colin, 1966) quoted in Wallerstein, *The Modern World-System I*, 45, n. 109.

45. Wallerstein, *The Modern World-System I*, 45.

46. Wallerstein, *The Modern World-System II*, 101.

47. Wallerstein, *The Modern World-System I*, 45.

48. Wallerstein, *The Modern World-System II*, 75.

49. Wocjiech Szcygielski, "Die Okonomische Aktivitat de Polnischen Adels im 16.-18. Jarhumdert," in *Studia Historia Economicae* 2 (1967), 94, quoted in Wallerstein, 1980, 133, n. 16.)

50. Wallerstein, *The Modern World-System II*, 161-162.

51. In the first volume of *The Modern World-System*, the only really curious omission in this discussion of monoculture and the environment is a failure to discuss the ecology of wheat, the other great export crop of the sixteenth century. While certainly not exhaustive of soil fertility to the same degree as sugar, we might observe, with Braudel, that "wheat's unpardonable fault was its low yield; it did not provide for its people adequately" (*Structures of Everyday Life*, 120) and that it "devours the soil and forces it to rest regularly," an ecological reality that "implied and permitted the raising of livestock" (Braudel, *Afterthoughts on Material Civilization and Capitalism* [Baltimore, Md.: The Johns Hopkins University Press], 11) — itself a major factor in ecological transformation as Wallerstein recognizes (op cit., 56-57, 63), and one that sets Europe's environmental history apart from, inter alia, China's rice agronomy.

52. Wallerstein, *The Modern World-System I*, 102.

53. Wallerstein, *The Modern World-System II*, 162, 162, n. 168.

54. Pierre Chaunu, "Bresil et l'Atlantique au XVIIe Siecle," in *Annales E.S.C.* 17, (6, 1961): 1176-1207.

55. Wallerstein, *The Modern World-System II*, 162.

56. Wallerstein, *The Modern World-System I*, 89.

57. Wallerstein, *The Modern World-System I*, 44; Wallerstein, *The Modern World-System II*, 162, n. 164.

58. Wallerstein, *The Modern World-System II*, 132-133, emphasis added; internal quotation from Szcygielski, "Die Okonomische Aktivitat de Polnischen," 94.

59. Karl Marx, *Capital*, Vol. I (New York: Vintage, 1977), 283, 636-638.

60. In terms of Marx's political economy, this line of thinking is developed furthest by Paul Burkett, *Marx and Nature: A Red and Green Perspective* (New York: St. Martin's, 1999). Moore, "Nature and the Transition from Feudalism to Capitalism," develops these themes historically.

61. Historical studies of consumption relations as a socio-physical moment of capital accumulation are not well developed. But see Raymond Grew, ed., *Food in Global History* (Boulder, Colo.: Westview Press, 2000); and Sidney W. Mintz, *Sweetness and Power* (New York: Penguin, 1985). Food consumption provides a particularly interesting angle of vision upon the dialectical relations between bodily space and the socio-spatial division of labor. The rise of "food consumption politics" as an emergent social movement—around such issues as genetically modified foods and recombinant bovine growth hormone—will undoubtedly spur new research into consumption relations as an aspect of world environmental history.

62. Wallerstein, *The Modern World-System I*, 90.

63. Wallerstein, *The Modern World-System I*, 36.

64. Wallerstein, *The Modern World-System I*, 44.

65. Karl Helleiner, "The Population of Europe from the Black Death to the Eve of the Vital Revolution," in *The Cambridge Economic History of Europe IV*, eds. E.E Rich and C.H. Wilson (Cambridge: Cambridge University Press, 1967), 68-69, quoted in Wallerstein, *The Modern World-System I*, 36, n. 78.

66. Wallerstein, *The Modern World-System I*, 109.

67. David Harvey, *The Limits to Capital* (London: Verso, 1999[1982]); Neil Brenner, "Beyond state-centrism? Space, territoriality, and geographical scale in globalization studies," in *Theory and Society* 28 (1999), 39-78.

68. On the matter of declining population density in the early modern Atlantic World, see Walter Prescott Webb, *The Great Frontier* (Austin: University of Texas Press, 1964).

69. Most recently, McNeill, *Something New Under the Sun*.

70. Worster, "World Without Borders"; Worster, "Transformations of the Earth: Toward an Agroecological Perspective in History," in *Journal of American History* 76 (4, March, 1990): 1087-1106; Hughes, "Global Dimensions of Environmental History," 101.

71. Robert Brenner, "Agrarian class structure and economic development in pre-industrial Europe," in *The Brenner Debate*, eds. T.H. Aston and C.H.E Philpin (Cambridge: Cambridge University Press, 1985), 10-62.

72. Fernand Braudel, *The Perspective of the World* (New York: Harper and Row, 1984); Giovanni Arrighi, *The Long Twentieth Century: Money, Power, and the Origins of Our Times* (London: Verso, 1994).

73. Wallerstein's agrarian geographical conception of capitalism is not without its problems, from the perspective of environmental history. Because capitalism creates a historically unprecedented antagonism between town and country, whereby the countryside is progressively "urbanized," the tension between urban power and agrarian landscapes in especially sharp, even in the early modern era (Karl Marx, *The Grundrisse* [New York: Vintage, 1973], 479; Moore, "Nature and the Transition from Feudalism to Capitalism"). Neglecting the rural-urban dialectic, then, obscures from view a significant vector of socio-ecological transformation in the modern era.

In my view, following Marx and Engels, capitalism's geography should be thought of as neither "urban" nor "agrarian" but rather as constituted by an evolving town-country division of labor. The town-country division of labor—not urban capitalism, not agrarian capitalism, and not some abstract "space of flows"—constitutes the overarching eco-geographical dialectic of the capitalist world-economy, through which the world-historical relations of class, capital, and metabolism flow. The rural-urban dialectic is something much more far reaching than a place-specific city-hinterland relation. Rather, the antagonistic town-country division refers not merely to place but to *space*, to those socio-spatial relations whose contradictions compel capital to produce new spatial configurations of accumulation and production and destroy the old. From this perspective, the rural-urban dialectic operates at multiple geographical scales (region, nation-state, world). The importance of this conception becomes especially clear in the light of capitalism's unsustainable and unequal ecological relations between country and city—a "metabolic rift" that compels capital to perpetually look for ways to achieve a "spatial fix" (Harvey, 1999) to its recurrent and ever-deepening accumulation crises. See John Bellamy Foster, *Marx's Ecology* (New York: Monthly Review Press, 2000); and Moore, "Environmental Crises." A critical assessment of the Wallerstein's agrarian conception of capitalism may be found in Jason W. Moore, "*The Modern World-System* as Environmental History? Transformations of Capital and Nature over the *Longue Duree*," paper presented to the 25th annual conference of the Political Economy of the World-System section, American Sociological Association, Blacksburg, Va., April 17-19, 2001.

74. Moore, "Sugar and the Expansion of the Early Modern World-Economy"; Moore, "Nature and the Transition from Feudalism to Capitalism"; Moore, "*The Modern World-System* as Environmental History?"

75. Wallerstein, *The Modern World-System I*, 121-122.

76. Foster, *Marx's Ecology*; also Moore, op cit.

77. Sauer, *Selected essays*, 360.

78. See, especially, Moore, "Environmental Crises," and "Nature and the Transition from Feudalism to Capitalism."

79. For instance, most recently, John Broich, "The Wasting of Wolin: Environmental Factors in the Downfall of a Baltic Town," in *Environment and History* 7 (2001): 187-199.

Chapter 3

Rousseau in the Suburbs: Geography, Environment, and the Philosophical Tradition

Robert Kirkman

The act of framing can sometimes do violence to that which enframed, cropping off inconvenient complexities and imposing boundaries that may be too narrow, too straight, or too rigid. This is an essay about the use of the philosophical tradition to frame human life and human practices within particular landscapes, especially a potentially violent mode of framing I call "selective philosophy." It is also an essay about the meaning of two intertwining strands of contemporary American life and thought: suburbanism and environmentalism.

The essay opens with a deliberately provocative act of framing: a reading of Rousseau as ancestor to and inspiration for American suburbanism. Through a series of juxtapositions—one frame against another—I hope to demonstrate that the mode of selective philosophy as such is suspect: the complexity of that which is enframed strains against the narrow set of meanings imposed upon it; the boundaries fracture, shift, and blur, suggesting that some more nuanced and multi-layered framing is needed to comprehend suburbia, environmentalism, and the connections between them. One way to begin would be with a more sophisticated reading of the philosophical tradition, a reading that leads to some surprising conclusions.

Rousseau as Suburbanite

I propose Jean-Jacques Rousseau as the first philosopher of suburbia. In his life and in his writings he is both inspiration for and ancestor to the patterns of suburban life that have come to dominate the United States since the nineteenth century, patterns that have been taken up and repeated around the world.

Modern-day suburbanites may look to Rousseau for validation of their own rejection of life in the city. As a young man Rousseau lived for a time in Chambéry with his protector, Mme. de Warens, in a house Rousseau found ugly and depressing. Although some aspects of town life appealed to him, especially as it offered him the opportunity to work as a music teacher, he continually sought some means of escape. For several years, Rousseau could take refuge in a cottage Mme. de Warens rented in a small walled garden in the suburbs. These were not the spacious and gracious suburbs that would appear in England and the United States in the nineteenth century, however, but the unfashionable and forgotten fringes of an Alpine town; the garden was crowded in by houses. Always in a financial crisis, Mme. de Warens was forced to give up even this meager sample of country life, much to Rousseau's dismay. In 1736, however, Rousseau convinced Mme. de Warens that they should spend their summers further out in the country. They settled at Les Charmettes, a "retreat far enough away for us to live in peace, and near enough for us to return at any time it may be necessary." This semi-rural setting was the backdrop for what Rousseau would come to call "the short period of my life's happiness."[1]

Two decades later, 1756, Rousseau left Paris to take up residence at the Hermitage in Montmorency, provided for him rent-free by Mme. d'Épinay. By that time, he was better able to articulate the cause of his discontent with urban life, and the problem lay not only with the physical environment but with the moral, social and political environment. The ugliness of Mme. de Warens's house in Chambéry was exceptional; more poisonous and more pervasive was the ugliness of human interaction, driven always by inscrutable and unnatural motives, and resulting always in the arbitrary power of some people over others. In the *Discourse on Inequality*, Rousseau had admonished anyone who could "to leave behind in the Cities your fatal acquisitions, your restless minds, your corrupted hearts, and your unbridled desires; resume your ancient and first innocence since it is in your power to do so; go into the woods to lose sight and memory of your contemporaries' crimes."[2]

Rousseau himself knew he could not resume his first innocence, but he was drawn away from the city nonetheless. Rousseau writes that, while he lived in Paris, "in the whirl of high society, at luxurious suppers, amidst the glitter of the theatre, in a cloud of vainglory," he found himself distracted and saddened by his memories of Les Charmettes. He describes his move to the Hermitage as a return to his own element: "I felt I was born for retirement in the country; it was impossible for me to live happily anywhere else."[3]

Even so, Rousseau was once again on the edge of the city, far enough away to be free of its corruption, but close enough that he could take advantage of its cultural and commercial resources. In 1756, Rousseau still earned his living as a music copyist, which made him directly dependent upon the cultural life of Paris, and none of his writings would have seen the light of day were it not for urban publishing houses. He also wanted the city nearby so that his friends (while he still had them) could come to visit, though few seemed willing to walk the five miles of bad roads from Saint-Denis after the long coach ride out from the city.[4]

Suburbanites may also see in Rousseau's life and work a reflection of their own preference for rural or semi-rural settings, landscapes that take a position in the middle between the city and the wilderness. The valley of Les Charmettes was, for Rousseau, the realization of the pastoral ideal. "Between two fairly high hills," he writes in the *Confessions*, "lies a little valley running north and south, down which a stream trickles over pebbles and under trees. Along this valley, halfway up the hill, are some scattered houses, very attractive to anyone liking a sanctuary somewhat wild and aloof." Clearly, however, the valley could only be described as "wild" in the relative sense that it was not urban. Rousseau describes the house he occupied with Mme. de Warens as "very habitable. In front was a terraced garden; above it a vineyard, below it an orchard, facing it was a little chestnut plantation, close by was a spring, and higher up the mountain were meadows for grazing cattle. In fact, here was everything that was necessary for the little country household we wished to set up."[5] In other words, this was hardly a wilderness encampment.

The fact of the matter is that the pastoral ideal is as much threatened by true wilderness as it is by civilization.[6] In the seventh walk of his *Reveries of the Solitary Walker*, Rousseau recounts a botanical adventure he had in the Alps during the first year of his exile. Wandering the slopes of a mountain called the Robeila, he worked his way "from thicket to thicket and rock to rock" to "a corner so deeply hidden away that I do not think I have ever seen a wilder spot." Rousseau was overwhelmed by its strangeness: the tangled wall of huge trees, the fearful precipices, and the cries of raptors. Only the calls of more familiar birds and the identification of familiar plants "lightened the horror" of the place. Once he had settled down to botanizing, Rousseau began to dream that he was "a second Columbus," discovering another new world. His reverie was interrupted by one more familiar sound: the clicking of a stocking mill, just twenty yards from *terra incognita*. This discovery elicited a contradictory response from Rousseau, but it is telling that his first reaction "was one of joy at finding myself among human beings where I had thought I was quite alone."[7] Raw wilderness was too much for him, it seems; it took a modicum of tame nature and human activity to put him at ease.

Remembering this episode leads Rousseau to reflect on the Swiss countryside as a whole, and he provides as good a description of the pastoral middle landscape as can be imagined:

> Switzerland is the one country in the world where you can find this mixture of wild nature and human industry. The whole of Switzerland is like one great city, whose long wide streets, longer than the Rue Saint-Antoine, are planted with forests and broken up by mountains, and whose scattered and isolated houses are only connected with one another by English landscape gardens.[8]

This same description could just as easily apply, on a smaller scale, to American railroad suburbs of the nineteenth century, or to the exurban landscape of five-acre zoning in the twenty-first.

Finally, Rousseau bolsters the modern suburban emphasis on privacy and leisure. As a matter of principle, Rousseau held that freedom from the arbitrary demands of others is essential to happiness. In practice, he sought independence from other people whenever he could get it, usually by means of physical isolation. In the *Confessions*, he makes much of the fact that Les Charmettes was "solitary," "aloof," and far enough from Chambéry "to throw importunate visitors off the scent."[9] The Hermitage was likewise solitary, although his happiness there was fatally poisoned by the fact that he owed a debt of gratitude to Mme. d'Épinay, who did not charge rent but made polite demands upon his time.[10] The closest he ever came to regaining the happiness of his idyll at Les Charmettes was, ironically, at the beginning of his exile. For six weeks in the summer of 1765, Rousseau found refuge on the Île de Saint-Pierre in the Lake of Bienne. There, physically cut of from those who would persecute him, Rousseau was able to live very much as he pleased in a rural setting he found delightful; he spent long days wandering around the island, falling into reveries, and occasionally botanizing.[11] In the *Confessions* Rousseau claims that, while he was willing enough to associate with his partner, Thérèse, and with the family of the steward of the island, he did all he could to prevent contact with anyone else: he wrote letters only grudgingly, and it is said that on several occasions he made use of a trapdoor in his room to escape visitors.[12]

Rousseau not only inspires and confirms the patterns of suburban life in England and the United States, his writings contributed directly to the formation of those patterns. Rousseau's influence fed into the broader currents of romanticism, which spread very quickly to England and from there to the United States in the early nineteenth century. In America, romantic anti-urbanism and pastoralism intertwined with the development of transportation and construction technologies in shaping the landscape, as railroads and balloon-frame houses brought semi-rural life within the reach of the middle classes. Romantic anti-urbanism was reinforced by the teachings of Catherine Beecher and other evangelicals who valorized the purity of a domestic sphere separated from commercial and political life. These sentiments were sharpened in reaction to the pollution, crowding, and increasing diversity of newly industrialized cities. The pastoral ideal of life in semi-rural, semi-wild landscapes came to be embodied in the writings of Emerson and Thoreau and in the paintings of Thomas Cole and the Hudson River School; romanticism combined with the aesthetics of the English countryside in the landscape designs of Olmsted and Vaux, including their design of Riverside, Illinois, the first fully-planned commuter suburb.[13]

Rousseau as Environmentalist . . . and as Anti-Environmentalist

There is, of course, something suspicious about any account that would cast Rousseau as arch-suburbanite. On the face of it, all I have done is to offer a narrow and selective reading of Rousseau's autobiographical writings in order to support a narrow and selective reading of the process of suburbanization. Even if Rousseau's thought had some influence on the early history of suburbanization, it would be neither accurate nor fair to say that suburbia is the pure expression of a romantic impulse that finds its sole source in Rousseau's writings. The modern American landscape has been shaped by a complex array of technological, economic, political and cultural forces, many of which may be seen as antithetical to ideal of a semi-rural retreat from civilization. Rousseau himself would almost certainly have been horrified—or simply baffled—by many aspects of twenty-first century freeway sprawl. If there is any truth to the stereotype of American postwar suburbs, then they are anything but a refuge from artificiality and from the arbitrary and intrusive power of other people. Americans are born free, Rousseau might observe, but everywhere they mow lawns.

I have made this proposal because its pattern should be familiar to philosophers who are engaged in public debates over land use and environmental quality: it embodies a particular way of using the philosophical tradition to frame some aspect of contemporary life, one that I call "selective philosophy." In truth, I first came up with the argument that Rousseau might serve as inspiration for and ancestor to "suburbanism" as a parody of the extant argument that he is inspiration for and ancestor to the environmental movement. Careful examination of the parallels between these two arguments, I believe, can yield some insight into the use and abuse of the philosophical tradition and into the peculiar character of environmentalism in the American landscape.

The extant argument for Rousseau as proto-environmentalist can be found in the work of Marcel Schneider and Gilbert LaFreniere, who have each appealed to Rousseau as both ancestor and inspiration to the environmental movement. LaFreniere maintains, for example, that "in addition to his contribution to the deistic nature worship which formed a powerful current in the Romantic Movement," Rousseau "is recognized as an important defender of nature for its value to humanity." In fact, LaFreniere continues, Rousseau is "the equivalent of an environmental radical," who calls for "the virtual reconstruction of society and of the economic, political and ethical attitudes and institutions which bear upon our relationship to the natural world."[14]

As a possible inspiration for environmentalism, LaFreniere maintains that Rousseau offers nothing less than "a model of a complete, holistic outlook toward nature and humanity's relationship to nature," including an implicit environmental ethic. Rousseau criticized his contemporaries' idea of progress, advocated pastoralism, fostered the deistic worship of nature, created a new aesthetic for the appreciation of nature, and even espoused a "transformist" account of human development which points "towards the evolutionary paradigm essential to modern ecology."[15]

This echoes the earlier work of Marcel Schneider, who placed special emphasis on Rousseau's notion of the state of nature as a basis for Green politics. Schneider notes, for example, that the principal publication of the ecology movement in France was called *Le Sauvage*. With his appeal to the noble savage, Schneider believes, Rousseau founded a new religion, one without dogma, "which teaches us to feel God in nature." One can be cleansed of the sins of society by becoming *sauvage*, by detaching oneself from "useless luxuries and pernicious comforts, from machines, and from technology and science." Stripped of all this, one can be filled and transported by the contemplation of nature "and see God in the ferocity of the sun, in the passing clouds, in the rain that falls, sweet and pure."[16] For Schneider, the ethical, social, and political payoff of this new religion is obvious: the state of nature becomes a model of utopia, an ideal toward which humans ought to strive.[17]

LaFreniere argues that Rousseau's role as an ancestor to modern environmentalism has been overlooked for a number of reasons, including various misrepresentations of his thought arising from the French Revolution and the difficulty of interpreting his critique of progress. Nevertheless, LaFreniere insists, the connection is there for anyone who cares to look for it. It has long been accepted that contemporary environmentalism can trace its roots to the Transcendental Movement of the nineteenth century, but the European roots of Transcendentalism are less often emphasized. LaFreniere cites documentary evidence that Emerson read the work of Rousseau and was deeply moved by it. By itself, this is a tenuous link. All of the Transcendentalists, however, were influenced by Wordsworth who in his youth was deeply influenced by Rousseau's writings. During his early travels in Europe, for example, Wordsworth made a point of visiting the Île de Saint-Pierre in the Lake of Bienne.[18]

This argument for Rousseau as proto-environmentalist seems plausible enough, more plausible perhaps than the argument for Rousseau as *proto*-suburbanite. Even so, there are difficulties with the arguments put forward by Schneider and LaFreniere, including the fact that Rousseau's ideas have just as plausibly been appropriated to place a very different frame around environmental politics. For Luc Ferry, Rousseau stands at the head of a humanist tradition that hinges on the belief that the essence of humanity resides not so much in rationality as in freedom. Elaborating on this, Ferry echoes Sartre as much as Rousseau: humans are anti-natural, which is to say that we are able to transcend any merely natural or social determinations. Coupled with freedom is Rousseau's belief in the perfectibility of humanity, which for Ferry implies the possibility of history, unfolding on two levels: "education as the history of the individual and politics as the history of the species." In sum, Ferry insists that human civilization cannot and should not be fragmented into particular local or national traditions to which individuals belong by nature; it is only by progressively detaching from such determinations that the world of culture, or mind, is constituted.[19]

Ferry uses this reading most directly against the animal liberation movement. Quite simply, animals cannot be said to have rights because they are not free, which is to say that they are incapable of transcending natural determinations. Ferry brushes aside the utilitarian arguments offered by Singer and others as incompatible with the liberal humanism of Rousseau and Kant. For his more general argument against environmentalism, Ferry takes a somewhat different tack: he characterizes environmentalism as a "fundamentalist" political movement that threatens to bring the era of humanism to a close by binding humans up once more in natural and social determinations.[20]

It is obvious that Ferry invokes Rousseau against a very narrow conception of what environmentalism is or should be: he lumps together a wide range of ideas, platforms and modes of activism under the name of "deep ecology," which he then dismisses with a single stroke. In the end, Ferry comes to advocate a different but still very narrow version of liberal environmental politics, characterized by competition among materialist interest groups. What may not be so obvious is that Schneider and LaFreniere invoke Rousseau in support of a conception of environmentalism that is just as narrow. Assuming a strict division between utilitarian conservationism and romantic preservationism, Schneider and LaFreniere clearly choose the latter: for them, environmentalism is a matter of defending pristine and providential nature against various kinds of technological intervention. It is hardly surprising that Rousseau's pastoral imagery resonates with such a project.

None of these narrow, too-tidy framings of environmentalism does justice to the diversity of the movement. In his recent study of pluralism and environmental justice, David Schlosberg goes so far as to say that "there is no such thing as environmentalism." The term is a convenience, "a vague label for an amazingly diverse array of ideas that have grown around the contemplation of the relationship between human beings and their surroundings." Schlosberg goes on to provide an impressive list of possibilities:

> the environmental field includes not only romantic preservationists, efficient conservationists, public-health advocates, and environmental illness victims, but also deep ecologists, greens, bioregionalists, animal liberationists, advocates of permaculture and organic agriculture, ecofeminists, religious evangelists, social ecologists, steady-state economists, neo-Malthusians, neo-Luddites, neo-Hobbesians, ecological technology promoters, nature consumers, indigenous rights activists, spiritualists, planners, conservation biologists, environmental health professionals, environmental justice advocates, environmental lawyers, gaians, ecosocialists, nature writers, worker-health advocates, eco-anarchist youth, and more.[21]

Rousseau might indeed be ancestor or inspiration to one or another of these perspectives. It would be misleading however to say that he therefore lends his support to environmentalism as such, which casts doubt on Schneider and LaFreniere's very broad claims on Rousseau's behalf.

At the same time, Rousseau's thought is itself complex and subject to a range of often contradictory interpretations; his ideas have been appropriated by—or foisted upon—every major political ideology that has arisen since the middle of the eighteenth century. Maurice Cranston states the matter this way:

> In the more than two hundred years since Rousseau's death, readers have disagreed sharply in their understanding of what he intended to say. Liberals, socialists, conservatives, anarchists, even the "Greens" of recent years, have all found footing on his literary landscape, either on the crags or in the crevices. He himself admitted there were contradictions in his writings; but he felt that what he had to say could not be expressed according to the rules of clear and witty exposition as practised by the philosophes. Rousseau had complex attitudes to what he considered, not unjustifiably, to be complex problems.[22]

Cranston points out two broad directions in which Rousseau's thought has been taken. On the one hand, Rousseau helped to foster the Romantic tradition, which emphasized sentiment over reason; "originality, spontaneity, and diversity over convention, artifice and conformity; the innocence and purity of the self against the corruptions of society; nostalgia for a natural paradise instead of confidence in progress and perfectibility." LaFreniere and Schneider have inherited this tendency. On the other hand, Rousseau helped to foster efforts to secure "civil freedom by political reforms based upon the republican principles of the social contract but taking account of particular conditions."[23] This is clearly the direction of Ferry's thought.

The readings of Rousseau as friend and as foe of environmentalism are particularly striking examples of selective philosophy: it is possible to pick and choose aspects and ideas from the philosophical tradition to suit any number of arguments, even advocacy of the suburban ideal. This comes at a high price, however. Selective philosophy does violence to the philosophical tradition and, once exposed, undermines the credibility of the arguments in which it is employed. The best that can be said of selective philosophy is that it offers a partial glimpse of a half truth: Schneider, LaFreniere, and Ferry are not exactly wrong in their readings of Rousseau, and their arguments may be illuminating to some degree, but in telling only part of the story they may be passing over important insights into the human condition.

The Uses of Philosophy

The search for understanding of our current environmental predicament requires an approach to the philosophical tradition that is much more cautious and critical than that embodied in selective philosophy. Philosophical insights into the human condition are not scattered around like pebbles on a beach to be collected or rejected at will. It might be better to think of them instead as strands woven into a dense fabric of influence and inference, tension and opposition; each concept, each model, extends or limits the meaning of others and is extended or limited by others in its turn. To learn much of anything from the philosophical tradition demands attention to history, context, and the interplay of ideas. The risk, of course, is that we might learn something other than what we expect, or want, or need.

When Rousseau aspires to the pastoral ideal, for example, he is tapping into a literary tradition that dates back at least to ancient Rome. Virgil, too, valorized the middle landscape between the city and the wilderness. This tradition is not a single lineage, however. In his analysis of pastoralism in American thought, Leo Marx has distinguished two branches of the pastoral tradition, one "popular and sentimental" and the other "imaginative and complex." In the American context, Marx notes, flight from the city and the rise of the wilderness cult were motivated by sentimental pastoralism, an "inchoate" and "ill-defined feeling for nature."[24] Rousseau's vision is much closer to what Marx calls complex pastoralism, insisting on a delicate balance between the moral liberty that is the prerequisite of civilization and the natural liberty that is the hallmark of human innocence.[25]

According to Marx's account, Rousseau lived and wrote in the period just before the industrial revolution began to transform pastoralism by forcing upon it an acute awareness of the intrusion of history, the irreversibility of technological change, and the foredoomed fragility of the middle landscape.[26] This awareness came to be embodied in a literary trope, the interrupted idyll, in which contemplation of a natural scene is disturbed by the intrusion of technology. The incident on the slopes of the Robeila in which Rousseau's contemplation was interrupted by the noise of the stocking mill fits this pattern, but Rousseau does not seem to bring away from it the same sense of foreboding that Hawthorne or Thoreau would take away from their encounters with steam locomotives. Rousseau does suggest in his political writings that the balance between freedom and nature embodied in civil liberty is fragile and subject to irreversible decay, but it would be at least another century before commentators on American culture would be able to recognize the role of technological change in this undermining the pastoral ideal.[27]

It is worth noting that Rousseau's belief in the fragility of civil society and the inevitability of its corruption is likely to sit well neither with environmentalists like Schneider and LaFreniere nor with democratic humanists like Ferry. Each camp must, for reasons of its own, hope that the center will hold, that life in civil society or in the middle landscape, is tenable and sustainable in the long run. Here, at least, the philosophical tradition does not cooperate with political rhetoric; instead of lending unambiguous support to one platform or another, it raises a possibility that is both surprising and disquieting to nearly everyone. Rousseau may not turn out to be correct in his belief that civil society is fragile, but his arguments on behalf of this belief should certainly give us pause.

Perhaps equally surprising is the ambiguity of "nature" in Rousseau's thought. One of the primary motivations for Rousseau's work is his search for the basic conditions for human happiness. He found it, he believed, in a kind of freedom that consists not in doing what one wants, but in never being compelled to do what one does not want to do. In short, Rousseau believes that humans can be happy only if they avoid any dependence on others, or any subjection to their whims. The only guarantee of independence is contentment, which Rousseau describes as a proportion between desire and power: if I do not want more than I am able to get for myself, he reasons, then I will never be discontented, and I have no need to be dependent on others or to manipulate them in pursuit of my own, selfish interests.

Nature enters into Rousseau's account of happiness at several points and in several guises: it is the primeval world in which humans, guided only by natural sentiments, live lives of contentment and peace; it is the benevolent and providential mother who offers Rousseau refuge from his tormentors; it is the cultivated Swiss countryside at Les Charmettes; it is the stark and horrifying wilderness. In short, the concept of nature is protean, shifting its form and its meaning to fit the need of the moment; what does not change is Rousseau's hope for a life of contentment.

Rousseau's treatment of nature as, in effect, an accessory to happiness might have important consequences for environmental philosophy, especially recent efforts to develop an environmental ethic for the built environment. Such efforts have shown a tendency toward what I would call a monovalent landscape criticism, one in which naturalness is taken to be the only relevant standard by which to assess patterns of human life in a given environment—a single frame to fit all circumstances. Within the literature of environmental philosophy, the naturalness of the built environment can be defined in any number of ways, from ecological sustainability to aesthetic harmony. Roger J. H. King, to take just one recent example, proposes a standard that could well encompass several different conceptions of nature: we should "green" the built environment by allowing our immediate surroundings both to reflect and to shape a more benign relationship with nature; one of his explicit principles is that "the built environment should make nature a visible, palpable presence in daily human experience."[28]

Much as Rousseau seems to have affixed the term "natural" to any landscape in which human contentment is possible, one recent proposal for an ethic of the urban environment seems to index naturalness to human well-being more generally. Alistair Gunn points out that no human environment can ever be wholly natural in the sense of being untouched by human influences. However,

> to the extent that human beings are able to thrive in a given environment—to live long lives, to be physically and psychologically healthy, to fulfill a significant portion of their potential—then that environment is natural for humans. Because we are so adaptable, there are a wide variety of environments that are natural for humans in this sense.[29]

Gunn calls such environments "natural," but he might just as easily have offered this as the definition of a "healthy" environment, or one that is simply "good," if only in the sense that it is good for the humans who inhabit it.

If my reading of Rousseau is correct, then it may be that the notion of naturalness is simply too protean to be of much use as a standard of landscape criticism. I suspect that a broader survey of the philosophy of nature would yield the same conclusion. Following Gunn, I would suggest that the most pertinent question to ask about any inhabited landscape is not whether it is "natural" but whether human beings can thrive there given the affordances and limitations of the surrounding environment.

This is by no means an easy question to answer. The description and prescription of the best kind of life for human beings has been a central problem for philosophical ethics for millennia. There are now and perhaps always will be competing notions of the highest good—another matter that may be clarified by careful and critical attention to the philosophical tradition. Moreover, landscapes themselves may be complex, serving the best interests of some while thwarting those of others, and always subject to widely divergent interpretations.

Such considerations have led me to the conclusion that landscape criticism should be multivalent, including considerations of sustainability and justice alongside those of happiness and aesthetic quality. The goal should not be to arrive at a single, unambiguous judgment about human patterns of life in a given environment, but to draw attention to the conflict and confluence of human values and aspirations as they play themselves out within the limits imposed by the landscape itself. This may not be what many environmental philosophers have hoped for, but I believe it is what our tradition and our training equips us to do.

Environmentalists in the Suburbs

I have said that the mode of framing typified by selective philosophy sometimes yields partial insights into half-truths. It may be that the juxtaposition of several frames produced by competing instances of selective philosophy might yield a set of overlapping partial insights into a set of overlapping half-truths. The result may not add up to truth, but it might at least suggest an interesting line of inquiry. My own juxtaposition of arguments on behalf of suburbia with arguments on behalf of environmentalism suggests that suburbia and environmentalism have something in common: both have at least one root in the pastoral vision of a finely-tuned balance between nature and culture. This connection is worth exploring simply because it might shed some light on the history of environmentalism and on the pastoral assumptions and aspirations that might linger in environmental theory. Moreover, it serves to complicate any environmentalist critique of suburban patterns of life.

I would place two caveats on this exploration, however. First, it is important to avoid giving a reductive account: I have already indicated that both suburbia and environmentalism are much too complex for either to be explained entirely in terms of the pastoral vision. Second, it is important to avoid imputing guilt by association, in either direction. Many environmentalists might feel threatened by the implication that they have anything at all in common with the suburban ideal, since many of them perceive suburbanization as an ongoing environmental disaster. At the same time, at least some suburbanites might be offended by the implication that they have anything at all in common with environmentalism, since some of them perceive environmentalism as an attack on civilization from the side of wildness. It should be clear from the beginning that the goal of any investigation into the parallel between environmentalism and suburbia is to understand the history of ideas and their intertwining with particular landscapes, not to insult people or to score political points.

The influence of the pastoral ideal on suburban patterns of living is most obvious in the earliest phases of suburbanization in the United States. In the nineteenth century, as the elite fled the increasingly crowded and polluted industrial city, they embarked upon what Robert A. M. Stern calls "the American search for an ideal community in nature." They hoped to preserve preindustrial values by conjoining two older patterns of living: the artisan's village and the open countryside. According to Stern, "the merger of these two patterns had the positive effect of creating a rich and visually varied environment comparable to the texture of the surrounded urban areas."[30]

Early suburbs tended to be large in scale and, because they were essentially new, they required the work of professional planners. These planners, including Alexander Jackson Davis, Frederick Law Olmstead, and Calvert Vaux, "for the most part employed naturalistic plans and vernacular architecture to create the illusion of country villages set in picture-perfect landscapes that had developed organically over long periods of time."[31] Davis' 1859 plan for Llewellyn Park, New Jersey, and Olmsted and Vaux's 1869 plan for Riverside, Illinois, set the standard for the first phase of suburbanization in America, with winding roads, houses widely separated by trees and park-like lawns, and common park lands such as the 50-acre Ramble at Llewellyn Park.

The pastoral ideal was not the only influence on suburbanization, of course. Other cultural forces were at work, as were a range of technological and economic forces. By the time the first Levittown was built in the late 1940s and early 1950s, suburbs had become both the product and the symbol of mass production and consumption. Nevertheless, Americans still hold on to the myth of the suburb as the middle landscape. On a recent visit to my family, I came across an article in the real estate section of the local newspaper singing the virtues of a small town a few miles to the west and celebrating its impending development into a full-fledged suburb of tract houses and strip malls. The headline read: "Rural relaxation, urban amenities."[32] Such a claim may well be nothing more than a cynical attempt on the part of developers to manipulate public opinion. The important point, though, is that the pastoral ideal of a semirural refuge has enough of a hold on the public imagination that it can be manipulated for commercial purposes.

What about environmentalism? Does the pastoral ideal still hold sway there? Avner de Shalit, for one, argues that it does not. In response to the claims of some historians that political ecology has its roots in right-wing ideology, de Shalit draws a careful distinction between ruralism and environmentalism within the ecology movement. Ruralism, like pastoralism, is "the glorification of country life, and a dissatisfaction with urbanism not only from the purely ecological point of view (e.g., as a source of pollution), but also because it is said to represent an inferior moral condition, or even a state of degeneration"; ruralism is romantic, antirational, and preindustrial in its orientation. By focusing entirely on the ruralist aspect of the ecology movement, de Shalit argues, Anna Bramwell and other historians have been led to the false conclusion that ecology is allied with fascism. What these historians have missed, de Shalit maintains, is the environmentalist aspect of political ecology, especially as expressed in the work of environmental philosophers. According to de Shalit, environmentalism is "a philosophy based on anti-speciesism and respect for all organisms, for life, for ecosystems, and so forth"; it is progressive and "rational, deriving from an empirical and scientific examination of the state of the environment."[33]

De Shalit is surely justified in protesting the exclusive focus on ruralism and its implied guilt by association. In the American context, it is certainly the case that the environmental movement has come a long way from the romantic wilderness cult of the late nineteenth century. The increasing complexity and vulnerability of technological civilization have shifted at least some of the impetus of environmentalism from the search for harmony with nature to the political struggle to preserve ecosystem integrity and human well-being in the face of direct physical threats. But even as Bramwell and others overstate the importance of romanticism in the modern environmental movement, de Shalit may be overstating the importance of the distinction between ruralism and environmentalism. He presents a stark choice between the two: one is either trying to "be green" in one's personal or spiritual life in order to bring about "the abolition of the rationalistic, bourgeois system in politics, and the creation of a new mentality," or one is engaging in "grassroots democracy and working within the existing political system" in order to bring about "a more environment-friendly society."[34]

This strikes me as a false dichotomy: not only are there other live options for environmental theory and practice, but it would seem that ruralism and environmentalism are not mutually exclusive. At least in the United States, concerns over pollution and resource depletion have not displaced love of wild and rural landscapes or the search for personal transformation; the two sets of concerns have intertwined in complex ways, further augmenting the diversity of the environmental movement. Consider, for example, the various initiatives enacted by the State of Maryland since 1997 under the banner of "smart growth." By and large, these are hard-headed proposals to reform political, regulatory, and market forces in order to reign in urban sprawl and redirect development into what some consider to be more benign pathways. The public justification for these initiatives is an odd mix of what de Shalit would call "environmentalist" and "ruralist" aspirations: *former* Governor Glendening has asserted that the people of Maryland want "green pastures and lush forests, strong and friendly neighborhoods, prosperous businesses and family-supporting jobs, good schools and a healthy environment." Alongside concerns about saving taxpayer money and making regulatory processes more fair and cost effective is the desire to preserve open space, natural beauty, and distinctive communities with a strong sense of place—including farming communities.[35] This explicit appeal to green pastures and rural villages does not, of course, amount to a reactionary ruralism that would call for the abolition of the current political order; it does suggest, however, that pastoral ideals and images have enough of a hold on the public imagination that they can be drawn upon to gain support for environmentalist policies.

So far, I have established nothing more than a correlation: some aspects of both suburbia and environmentalism incorporate some elements of the pastoral vision. There is evidence, however, of some mutual influence between the two, especially in that some aspects of contemporary environmentalism have their origins in the suburbs. Christopher Sellers has examined the case of DDT spraying on suburban Long Island and argues that two lawsuits—the first in 1958, the second in 1966—helped to define environmentalism as a new political identity. The first case was brought by wealthy landowners on the elite North Shore of Long Island, seeking to protect their property and their own bodies from the perceived dangers of DDT. Although they lost their case, their effort did not go unnoticed. Rachel Carson's *Silent Spring* was shaped in part by the 1958 trial, through Carson's contact with Marjorie Spock, one of the plaintiffs. The second trial was brought by residents of newer suburbs in Suffolk County, and led to the establishment of both environmental law as a distinct field and the Environmental Defense Fund as a mainstream environmental organization.[36]

Ironically, at least some of the impetus for contemporary environmentalism in the suburbs came about as a reaction against the worst impacts of suburbanization, especially threats to clean water, clean air, and open space. According to Adam Rome, it was the destruction of open space that most galvanized suburbanites. Part of the appeal of life in the suburbs is easy access to undeveloped land. Rome cites a long passage from Margo Tupper's book, *No Place to Play*, which relates a suburban child's distress over the destruction of a nearby woodland to make way for another subdivision; it is because of experiences like these that the bulldozer became a powerful symbol of the destructive power of technology. Rome notes that the suburban reaction to the bulldozer was more intense than public reaction to conservation battles in the past, such as that over the damming of Hetch Hetchy in the early twentieth century, simply because the threat was more immediate: "The ravagers were not far away, building a massive structure in a valley of sacred beauty—they were next door, doing everyday work, and they threatened the well-being of families at home."[37]

Rome understates both the ambiguity of the bulldozer as a symbol and the irony of the suburbanites' response to it: it was the bulldozer, after all, that created the very arcadia that suburbanites were attempting to defend against the ravages of the bulldozer. What sense can be made of this? Perhaps this strand of American environmentalism is an expression of disappointment on the part of suburbanites that the suburbs are not suburban enough, which is to say that they do not live up to the pastoral ideal of a simple life in harmony with nature. Even if a new subdivision seems precisely balanced between the city and the wilderness, that balance is doomed to disruption as soon as the next subdivision is built and the congestion on the highway ratchets up another notch. There is a bit of dark humor that circulates among activists and planners concerned about the problems of sprawl: everyone wants to be the last to move to the suburbs.

There is obviously a great deal more to be said about the relationship between environmentalism and suburbia. It would be interesting, for example, to investigate the role of early conservationism in motivating people to leave the city, or the role of geography in shaping the constituency of mainstream environmentalist organizations. The account I have offered is complete enough, however, to establish that it might be worthwhile for environmental philosophers to examine the relationship in greater detail: as with careful study of the philosophical tradition, such an examination might teach us something surprising about ourselves.

Notes

1. Rousseau, *Confessions*, trans. J. M. Cohen (New York: Penguin, 1953), 214-215. See also Maurice Cranston, *Jean-Jacques: The Early Life and Work of Jean-Jacques Rousseau, 1712-1754* (Chicago: University of Chicago Press, 1982), 116-119.

2. Jean-Jacques Rousseau, *The First and Second Discourses*, ed. and trans. Victor Gourevitch (New York: Harper and Row, 1986), 213.

3. Rousseau, *Confessions*, 374-375. See also Cranston, *The Noble Savage: Jean-Jacques Rousseau, 1754-1762*, (Chicago: University of Chicago Press, 1991), 21.

4. Cranston, *The Noble Savage*, 23.

5. Rousseau, *Confessions*, 214.

6. See Leo Marx, *The Machine in the Garden: Technology and the Pastoral Ideal in America* (Oxford: Oxford University Press, 1964), 22.

7. Rousseau, *Reveries of the Solitary Walker*, trans. Peter France (New York: Penguin, 1979), 117-118

8. Rousseau, *Reveries of the Solitary Walker*, 119.

9. Rousseau, *Confessions*, 213-214.

10. Rousseau, *Confessions*, 382-384.

11. So powerful was his experience on the island that Rousseau wrote about it twice: *Confessions*, 588-596; and the "Fifth Walk" of *Reveries of the Solitary Walker*, 81-91.

12. See Maurice Cranston, *The Solitary Self: Jean-Jacques Rousseau in Exile and Adversity* (Chicago: The University of Chicago Press, 1997), 137. Cranston affirms the trapdoor story, but suggests that Rousseau exaggerated his own isolation; there is evidence that Rousseau in fact willingly received a constant stream of visitors.

13. See Kenneth T. Jackson, *Crabgrass Frontier* (Oxford: Oxford University Press, 1985), chapters 3 and 4.

14. Gilbert F. LaFreniere, "Rousseau and the European Roots of Environmentalism," *Environmental History Review* 14:4 (Winter 1990): 42.

15. LaFreniere, 42-43.

16. Marcel Schneider, *Jean-Jacques Rousseau et l'espoir écologiste* (Paris: Éditions Pygmalion, 1978), 25-26. Translation mine.

17. Schneider, 76. Regarding utopia, see also LaFreniere, 49.

18. LaFreniere, 60-62.

19. Luc Ferry, *The New Ecological Order*, trans. Carol Volk (Chicago: University of Chicago Press, 1995), 5-10.

20. Ferry, 60, 89-90.

21. David Schlosberg, *Environmental Justice and the New Pluralism* (Oxford: Oxford University Press, 1999), 3-4.

22. Cranston, *The Solitary Self*, 188.

23. Cranston, 191.

24. Marx, *The Machine in the Garden*, 5.

25. Marx, 101-102.

26. Marx, 362-365.

27. "Although a people can make itself free while it is still uncivilized, it cannot do so when its civil energies are worn out. Disturbances may then destroy a civil society without a revolution being able to restore it, so that as soon as the chains are broken, the state falls apart and exists no longer; then what is needed is a master, not a liberator. Free peoples, remember this maxim: liberty can be gained, but never *regained*." Jean-Jacques Rousseau, *The Social Contract*, trans. Maurice Cranston (New York: Penguin, 1968), 89.

28. Roger J. H. King, "Environmental Ethics and the Built Environment," *Environmental Ethics* 22 (2000): 115, 130.

29. Gunn, Alistair S., "Rethinking Communities: Environmental Ethics in an Urbanized World," *Environmental Ethics* 20 (1998): 348.

30. Robert A. M. Stern, *Pride of Place: Building the American Dream* (Boston and New York: Houghton Mifflin and American Heritage, 1986), 125

31. Stern, 126-127.

32. Mary-Beth McLaughlin, "Rural relaxation, urban amenities: Many developers consider Swanton northwest Ohio's next hot spot," *The Blade* (Toledo, Ohio) April 29, 2001, H1.

33. Avner de Shalit, "Ruralism or Environmentalism?," *Environmental Values* 5 (1996): 48, 50-51.

34. Shalit, 54.

35. Parris N. Glendening, "What is Smart Growth?: A Message from Governor Glendening," http://www.smartgrowth.state.md.us/whatis.htm, accessed January 18, 2002.

36. Christopher Sellers, "Body, Place and the State: The Makings of an 'Environmentalist' Imaginary in the Post-World War II U.S." *Radical History Review* 74 (1999): 31-32, 48.

37. Adam Rome, *The Bulldozer in the Countryside: Suburban Sprawl and the Rise of American Environmentalism* (Cambridge: Cambridge University Press, 2001), 150.

Chapter 4

Toward a Phenomenology of Cognitive Mapping

Gary Backhaus

Orientation: Returning to the Source and Identifying the Lacuna

In this chapter I frame an initial exploration toward a thorough phenomenological grounding of cognitive mapping. I take issue with the standard phenomenological corpus concerning the treatment of spatial experience, because its basis for establishing such a description of cognitive mapping has been left inadequately framed. I do not mean to take away from the important work and significant advances that phenomenologists have accomplished concerning the spatiality of lived-experience. However, I am critical of the emphasis on the structures of embodiment, which are articulated at the expense of the structures of the EarthBody to which they are in relation.[1] This bias results in the description of spatial experience that obscures the proper excellence of the phenomenological starting point: the primordial epistemological structure is the *relation* between the experiencer and experienced. Traditionally, the fundamentality of this relation involves the subject (noetic)–object (noematic) structure of intentional consciousness and the embodiment–environs structure of lived-body intentionality. The classical framing of this primordial epistemological relation views constitutive agency solely on the side of the intentional consciousness and bodily intentionality. So, the treatment of spatial experience by Edmund Husserl and subsequently by major phenomenologists does not adequately account for the constitutive function of what I will call Earthly-embodiment or Worldly-orientation. Worldhood is always a horizonal moment to the primary relations mentioned above and is an essential thematic structure in the higher levels of cognitive mapping. In this study I provide a fuller descriptive account of spatial orientation, which calls into question the adequacy of the emphasis solely placed on the lived-body's function as the zero point for the grounding of an orientational system of spatial coordinates. This function is propounded as the basic constitutive structure of qualitative spatial experience.[2] After its inaugural description by Husserl, this phenomenological structure is advanced in the work of various others such as Alfred Schutz, Maurice Merleau-Ponty, and Edward S. Casey, each of whom have enriched and have developed it in far-reaching ways.[3] However, the initial Husserlian analysis remains incomplete in terms of a thorough grounding of spatial orientation. Although the enrichments by others have indeed covered over its inadequacies, it yet needs to be reframed, for its fundamental structurings that fulfill the role of founding other structures should be well grounded. My goal is to return to Husserl's discussions in order to reframe a phenomenological description of the spatial moment of experience. My reframing takes into account this structure that I am calling worldly-orientation and earthly-embodiment, which opens the possibility of breaking new ground toward a phenomenological description, or framing, of cognitive mapping.

The goal of phenomenological description is to capture the essential structures comprising lived-experience. In contrast, the goal of explanatory science is to propose theoretical models that explain causal relationships. Phenomenology attempts to describe experiential content just as it appears without importing theoretical presuppositions that bias the description of experience. In the geography and psychology of spatial cognition one can find many theories concerning cognitive mapping. Through scientific experiment these theories are tested with the basic functional tools of protocols, controls, operations, etc. The empiricists' task projects a different goal than that of phenomenology's, which is to remain true to the experience and to uncover those material a priori structures necessary for the experience to be the kind of experience that it is.

I shall consult key passages in Edmund Husserl's *Ideas Pertaining to a Pure Phenomenology and to a Phenomenological Philosophy* Second Book (*Ideas II*), for it is there that he treats embodied consciousness, which allows him to differentiate structures constitutive of each of the regions of experience. In *Ideas I*, Husserl develops the notion that Being is not everywhere the same and formulates the task to set the parameters for various regional ontologies, each of which articulate a part of the whole of Being, which then is to be subject to the grounding discipline of transcendental phenomenology.[4] The following book, *Ideas II*, offers a static phenomenology that grounds the most fundamental structures constitutive of the major ontological regions—material nature, animal nature, and the spiritual world. In these texts Husserl is proceeding by what he calls the "Cartesian way" into phenomenology. He has introduced the ego into phenomenological analysis—the polarity of ego and object, which was absent in his act psychology promoted in the earlier *Logical Investigations*. In both *Ideas* cited, Husserl had not yet sufficiently developed a phenomenology in which *the world is co-constitutive*. I argue that worldly-orientation must be co-constitutive in its spatial-morphological sense (earthly-embodiment) in order for the experience of cognitive mapping to be phenomenologically described in an adequate way. The decisive point that I want to make is that Husserl had yet to ground a way into phenomenology that is commensurate to the needs of the description.

It must be understood that phenomenology can only maintain its integrity when *theoretical construction* does not slip into its description. In other words it must let experience present intuitive evidence and not let concepts speak for (construct) experience. For construction in this sense functions as theory that has not been properly presented for its operationalization in testing. At its worst, then, phenomenology can appear as subjective conjecture or "feelings." Phenomenological evidence must be given in intuition; its objects rigorously must be brought to experiential presence (even if present as absence). Husserl's work at the time of *Ideas II* was only capable of grounding a certain set of limited structures because of the strictures Husserl had to impose on the development of phenomenology in order to avoid constructivity. Nevertheless, the structures grounded in *Ideas II* are important to the project at hand, but the structures concerning the orientational system of embodiment need to be augmented through later breakthroughs in phenomenological modalities. In order to account for the experience of cognitive mapping, I employ a phenomenology that accounts for the co-constitutive function of the world. And, I also employ a genetic phenomenological methodology, which allows for the description of the layers of significance necessarily presupposed in the achievement of an integrated transcending orientational system.

In *Ideas II* the Body (*Lieb*, the animated organism) and its spatial orientation are given much attention but Husserl does not specifically articulate the structure of experience that the geographers call "a cognitive map"—a most necessary component of spatial experience. If one looks through *Ideas II* to find even a subsection dealing with cognitive mapping under the rubrics of the constitution of material nature, animal nature, or the spiritual world, no description is found. Given the central importance that spatial cognition, cognitive mapping, and cognitive maps have for geographical acumen, Husserl's lacuna is glaring. Since most geographers are empiricists, the study of cognitive mapping is then solely left up to theories of explanation and the anthropological examination of predilective differences in spatial cognitive styles. It is in the utmost interest for a phenomenological or lived-geography that cognitive mapping is given much attention.

Cognitive mapping has been developed in a variety of sophicticated ways in the discipline of geography.[5] I begin my inquiry with its the most basic spatial moments and genetically build its sense to more complex processes while still remaining within its most fundamental constitutive articulation. The particular sense of a cognitive map, which corroborates my understanding, has been defined as "long-term stored information about the relative location of objects and phenomena in the everyday physical environment."[6] Cognitive mapping develops into the ability to image in one's mind how one gets from one location to another. Regional schemata are developed such that one can move around within various "landscape districts" or regional parts with awareness of one's location in relation to other locations within each district. In addition one develops the ability to articulate the relation of the districts comprising the landscape and to imagine various points as points of reference such that perceivable parts of the landscape can be related to those points. Cognitive mapping involves a series of orientational developments the highest of which requires the integration of orientational systems into one transcending orientational system.

Explication of Key Passages and their Relevance to Cognitive Mapping

Relevant discussion in *Ideas II* is found in "Chapter Three: The Aestheta in Their Relation to the Aesthetic Body" of "Section One: The Constitution of Material Nature." Husserl states: "The Body is, in the first place, the *medium of all perception*; it is the *organ of perception* and is *necessarily* involved in all perception."[7] The discussion of this point further motivates examination of how the Body is related to things perceived. Husserl continues: "All that is thingly-real in the surrounding world of the Ego has its relation to the Body."[8] The Body's relation to the environs now becomes the leading question. "Connected with this is the distinction the Body acquires as the bearer of the zero point of orientation, the bearer of the here and now, out of which the pure Ego intuits space and the world of the senses. Thus each thing that appears has *eo ipso* an orienting relation to the Body, and this refers not only to what actually appears but to each thing that is supposed to be able to appear."[9]

In addition to this structure by which the Body relates perceivable objects to itself, which establishes the zero point of orientation, the Body "is of *significance for the construction of the spatial world.*"[10] This involves a corresponding relation of motivating and motivated systems of sensations. The motivating order involves a series of kinesthetic sensations, that is, the movement of the head or eye, that corresponds to sensations that carry the extensional apprehensions or images of the perceived object. If I keep my eyes fixed on a point and change my spatial location corresponding adumbrations of the object are motivated. "We constantly find here this two-fold articulation: kinesthetic sensations on the one side, the motivating; and the sensations of features on the other, the motivated."[11] It is this correspondence of the motivating with the motivated sensations that allows for objects to be organized around the Body in a "lawful" manner. The Body performs preconscious operations, spontaneous behaviors that can be counted on as delivering a fundamentally predictable style in the experiencing of its (in this case visually perceivable) environs.

Let us augment this discussion to examine fundamental cognitive mapping capacities. The first point is that the Body is not just a psychophysical unity, but also a "thingly-real" entity. It is of fundamental importance in terms of the qualitative experience of spatiality to recognize the Body as the zero point of orientation, but it acquires that assignation from the experiencer's point of view.

From the standpoint that the Body too is a "thingly-real" object, this assignment seems quite arbitrary: why this entity as "zero" rather than another? That we have the cognitive ability to recognize that the Body as a material entity--body (*Körper*, physical matter) is one amongst many material entities is fundamental to our ability to relate objects to other reference points as well. This ambiguity means that we can *alienate* the Body's orientational system by imaginatively placing it at another location and relate objects around that location. This possibility of constituting the Body as also a thingly-reality (body) allows the imagination the freedom to establish another thingly-reality as a zero point of reference, assigning a spatial orientation around it, and thus to imagine one's Body at its present location as a thingly-reality within the thingly-reality's orientation. The alienation of the Body's orientational system that constitutes the "Body" as a body in a visual field of reversed coordinates is one of the fundamental and spontaneous capacities involved in the act of cognitive mapping.[12]

A simple illustration will make this capacity clear. I see the tree at the far end of my property next to the mailbox. From my point of view both are relatively far from where I am standing, but from the standpoint of the mailbox the tree is close to it. Close to me is the holly bush, but from the standpoint of the mailbox the holly is relatively (in terms of the yard) far away. I can only make these observations on the basis of being able to transfer the original set of coordinates that is a spontaneously capacity constitutive of my own embodiment to the mailbox so that *assigned to it* are such qualitative dimensions as near/far. Of course this is an anthropogenic transfer of my own kind of experience, but it is the only manner of spatial experience that I can naturally promote—I am not a starfish or a bee.

In Bodily experience, the zero point of the set of coordinates takes on the specific spatial quality as *here*. "The Body then has, for its particular Ego, the unique distinction of bearing in itself the *zero point*. . . . One of its spatial points . . . is always characterized in the mode of the ultimate central here: that is, a here which has no other here outside of itself, in relation to which it would be a 'there.' It is thus that all things of the surrounding world possess an orientation to the Body."[13] The fundamental here-there structuring emerges on the basis that the Body is always here. But what is still in need of an acount is the *relationship between heres*. An inadequate account of the relation between heres is that on the basis of locomotion a former here becomes a there and a former there becomes a here, which sidesteps the question concerning the relation of the heres. In other words, what establishes a presently experienced *relation of* two zero reference points to one another, each with its own set of coordinates? The mere transpositions of theres into heres and vice versa are not subject to the problem of relating two separate systems. Husserl states: "The Body of the subject 'alters its position' in space; the things appearing in the environment are constantly oriented thereby; all appearances of things preserve the fixed system. . . . The form of the order of orientation around a center, all this is necessarily preserved."[14] Husserl is quite right about the orderly arrangement that is retained, but this aspect of experience that preserves things in a fixed system, provides a false sense of adequacy in terms of the question. Unless the Body retains a relationship with the former here as a here and is able to relate it to its present here, it would lose its previous spatial orientation. It would retain orientation to those objects through its present perceptions but would lose the prior orientation system in its own previous experience. The preserving of a fixed system or arrangement between perceptual elements is necessary but an insufficient condition. The question remains to be answered: what establishes the basis for the relationship between two different orientation systems?

Even though on the basis of the thingly-aspect (*Körper*) of my Body (*Lieb*) I can imagine some location as a zero point, and thus place my thingly-body in its orientational system, I cannot imagine that location as here, or "as a here" and still retain the here of my Body. The reason is that "here" only makes sense on the basis of the Body as a psychophysical unity and so I cannot displace it and retain it at the same time. If I imagine some other point as my here, I can certainly do so, but I cannot then place my Body as here in its field, because I then would experience two orientational systems simultaneously, which is impossible. I can only alternate between the orientation sytem of my Body's here or my Body's alienated orientational system. Or, if I stick to the sense that I imagine another zero point coordinate system, I remain with the fact that I (as psychophysical unity) am imagining it from here, *where I am*, and thus cannot think of that zero point as my here. This capacity constitutes *the reversibility of orientational systems*, the one or the other, but not both simultaneously.[15] This experiential limitation affirms the fundamental fact that I cannot be in two places at one time. If it is objected that having two orientational systems at once can be imagined, I respond that such an image is empty, in other words, it is analogous to conceiving a round-square.

But I can constitute another here on the basis of intersubjectivity. Husserl states, "My appearances belong to me, his to him. Only in the manner of appresence can I have, co-given with his Body, his appearances and his 'here,' to which they are related."[16] With the possibility of what Husserl labels as "analogizing" it is then possible for me to cognitively grasp two heres in my experience, mine and the other's, along with the two zero point orientational systems. Yet, they are in my consciousness in two different ways. My own, which is given intuitionally in perceptual acts and the other's, which I have on the basis of analogizing appresence—perceiving his animated Body and indirectly intuiting his consciousness. My here, for him, is there, yet as an Other's here, and his here for me is there, yet as an Other's here, and each of us empathically recognizes the intersubjectively spatialized bi-locality of our points of view—the *interchangeability* of our two qualitatively related standpoints.

Intersubjective experience allows me to perceive a here at some location different than the one I occupy and relate it to my own here. By establishing the two heres, I experience the *interchangeability of standpoints*. This ability to grasp two points of views in the same experience, and for that matter still other points of view, is another capacity that is a necessary component in the development of cognitive mapping. But we still must be careful to preserve the *re-*

versibility in these *interchangeable orientation systems*. I either refer to my own here and its perspective or to my partner's here and her perspective. I am aware of both orientation systems as being present at the same time; but I do not experience both perspectives at once, which again (as a heuristic analogy) would be like trying to synthesize a square-circle or a circle-square.

But a most important aspect of experience for cognitive mapping involves the ability to integrate heres into one system. This leads us to examine experience for a higher level capacity that achieves a transcending integration of the two (and a plethora of other) perspectives. Without this ability we would be unable to integrate the multifarious systems of oriented space in a way that frees us from our surroundings so that we can *image* the *totality*, an achievement of orientational transcendence necessary in the development of a cognitive map. Since we know that this capability exists, the next challenge is to examine how such an integrated and transcendent map is constituted in our experience.

Husserl's Abstraction: Objective Space

So far I have explicated Husserl's study in a way that shows its merits for a description of some of the essential structures inherent to the experience of cognitive mapping. At this juncture, I have introduced enough that the main critical point can be further explicated. I want to point out that Husserl fails to apprehend the structures constitutive for the transcendent integration of zero point orientation systems necessary for the further development in the genesis of cognitive mapping. Husserl's discussion leaps over the problem by objective abstraction: "But from that 'here' I can then consider even my Body as a natural Object, i.e., from that 'here' my Body is 'there,' just as the other's Body is 'there' from my 'here,' is there at a point in Objective space."[17] What I strongly object to in this passage is the notion of "Objective space." How is Objective space derived experientially? What Husserl means by "Objective space" (translators capitalize "Objective" in order to distinguish its specific meaning) is "intersubjective space." If we are not careful, we will think that on the basis of the interchangeability of standpoints in instersubjective space that we have constituted Objective space. If we say we have achieved intersubjective space, then the integrated space needed for higher levels of cognitive mapping has not yet been achieved. If we say that we have not achieved intersubjective space, it is because a genuinely transcendent integrated space has yet to be achieved. If what Husserl calls Objective space is not constituted in experience yet is meant by him, then Husserl is smuggling in a theoretical construct and breaking his own cardinal rule of phenomenological rigor. It must be recognized that all of the spacial capacities that have been discussed so far rest on the basis of the Body (what Husserl calls oriented spaces). The intellectualist notion of Objective space severs its relation to lived-space, about which Husserl should be wary, unless he wants to recant his position concerning the crisis of the European sciences. My claim must be demonstrated.

The following is his discussion.

> Every subject has his "space of orientation," his "here" and his possible "there," this "there" being determined according to the directional system of right-left, above-below, front-back. But the basic form of all identification of the intersubjective givenness of a sensuous content is of such a kind that they necessarily belong to one and the same *system of location*, whose Objectivity is manifest in that every "here" is identifiable with every relative "there" as regards every new "here" resulting from the subject's "moving on" and so also as regards every "here" from the viewpoint of another subject. This is an ideal necessity and constitutes an Objective system of location, one that does not allow of being grasped by the vision of the eyes but only by the understanding, that is, "visible," in a higher kind of intuition, founded on the change of location and empathy. In this way is solved the problem of the "form of intuition" and spatial intuition.[18]

How is the "Objective system of location constituted"? Not through sensuous intuition, but as an intellectual intuition founded on it. The act of naming Objective space and indicating its basis in intellectual intuition does not describe how such an intuition is constituted. But I argue that what Husserl should be after here is not Objective space as he means it. Objective space, for Husserl, turns out to be an ideal homogeneous coordinate system that reduces qualities to quantities, and such a system abstracts from the spatial character of experience—the space intuited in cognitive mapping. Husserl has the experiencer perform a highly intellectual theoretical act. "It comes to givenness by means of an identification with a change of orientation, but exclusively one the subject carries out freely. Oriented space (and along with it, *eo ipso* Objective space) and all appearing spatial forms already admit of idealization; they are to be grasped in geometric purity and determined 'exactly.'[19] This moment may be peculiar to those like Husserl who earned their doctorate in mathematics—it is an anthropological style or predilection. This is not the transcendent integrated orientation system that is formed through concrete experience of spatiality. Through the achievement of transcendence and integration in the developmental genesis of cognitive mapping we do not arrive at Objective space. Unless we are academic mathematicians, who perhaps do not leave their offices, we do not usually shift into a theoretical attitude by which mapping becomes an objective idealization. This bit of humor serves to remind the reader that this study is not guided by the empirical ways that people learn or employ cognitive mapping. Rather, a genetic phenomenology considers the order of the structures that build levels of significance. We arrive at a qualitative mapping that is filled in with the sensuosity of the landscape, which in terms of the genesis of sense is prior to, and differentiated from, an idealized template of theoretical consciousness. Kantians would reverse this notion, for they argue the intellectualist position that sensory existence is inchoate. Even though Husserl attempts to avoid Kant's formal a priori in his work, Husserl has to skip over the necessary experiential basis in sensuous intuition that allows for its idealization in a mathematical coordinate system. This important level in the genesis of sensuous intuition is missing because he fails to take into account the constitutive moment of worldly-orientation and earthly embodiment as co-constitutive for cognitive mapping.

To be fair to Husserl, the above quote is found in the first main section concerning the ontological region, material nature in which he is interested in constituting nature as a theoretical object of natural science. *But, theoretical objects are already constituted as pre-theoretical objects.*

> One must make clear to oneself that it belongs to the peculiar character of the theoretical attitude and its theoretical acts . . . that, in them, objects which for the first time *will become* theoretical are *already, in a certain manner, laid out there in advance.* Thus objects are already constituted pre-theoretically and are not Objects intended in the pre-eminent sense, and much less are they Objects of theoretically determining acts.[20]

This pre-theoretical/theoretical relation allows me to articulate Husserl's mistake. He has not sufficiently provided the pre-theoretical constitution of a transcendent and integrated oriented space from which what he calls Objective space (a theoretical abstraction) can be constituted through a theoretical act. Husserl unwittingly performs a theoretical act without the necessary pre-theoretical objects given to him in intuition. He himself makes the egregious mistake that he labels the crisis of the European sciences: constructing abstractions for their own sake that cannot be traced back to lived-experience and the lifeworld. To be more precise Husserl unwittingly forgets his own condemnation of Kant's formal a priori system and makes an intellectualist mistake. But it is necessary to investigate whether the pre-theoretical structure can be found in other major sections of Husserl's study.

In "The Constitution of the Spiritual World," subsection 50, "The person as the center of a surrounding world," Husserl restates his points at the level of spirit rather than Body: "As person, I am what I am . . . as *subject of a surrounding world.* The concepts of Ego and surrounding world are related to one another inseparably."[21] Even though Husserl grounds the inseparable relation of Ego and world in this work, he does not, from his standpoint of Cartesian egology, allow the world any constitutive role. "To begin with, the world is, in its *core*, a world appearing and characterized as 'on hand,' a world given in straightforward empirical intuitions."[22] Husserl does not mean by this that one perceives "the world," which is absurd, "it is the surrounding world only to the extent he 'knows' of it, insofar as he grasps it by apperception and positing or is conscious of it in the horizon of his existence as co-given."[23]

To recapitulate, my claim is that the more sophisticated levels of cognitive mapping cannot be phenomenologically described from the egological standpoint. It is necessary to promote a genetic phenomenology in which a subject (not merely an ego-pole) carries a history of sedimentations of spatial experiences that can be built up into "an atlas of maps." And, we must pay utmost attention to the Worldly-orientation and Earthly-embodiment, in other words, that the world is not merely a container for perceivable objects making up a landscape—that the World horizon exhibits a co-constitutive agency. In order to grasp how the necessary transcendent integration of orientational systems is achieved I embark on my own further phenomenological examination of spatial experience.

The Phenomenological Search for the Constitutive Structures Inherent to Cognitive Mapping

Every here entails a new coordinate system but of the same form. Here functions as the zero point of an orientational system of qualitative space surrounding the Body to which objects of perception are related. Every particular instantiation of a here is different in terms of objects-to-Body relations. The zero point and its surrounding zone constitutive of a particular here exhibits different object-relations than another here constituted as a zero point and its surrounding zone. Ambulatory locomotion is the agency of conversion. A here that has been converted into a there may or may not remain within the perceptual field. Nevertheless, on the basis of locomotion, the Body has performed the conversion. The here that is converted to a there is not merely a there among other theres, but is a *former-here-converted-into-a-there*. The former here is related experientially to the present here through the locomotive project. I start out from a here whereby objects are oriented perceptually around my Body. I then perambulate and come to rest whereby I take stock of my surroundings, meaning I constitute my present location as my here—as the zero point of my orientational system. I am limiting my use of the word "here" to entail conscious attention in the establishment of a here qua here, otherwise the experience remains in the horizons (non-thematic). *If* nothing is constitutive of the relation between the former here and the present here, there is no *connecting orientation* outside of the presently experienced here-there(s) organization. I may remember my past heres but there would be no connecting relation. Why? Since every constituted here has its own coordinate system, each one a zero point, I just have a series of unrelated systems. So, memory has to be able to recall the past orientation system in a way that relates and connects it to the present orientation system. Otherwise we would be oriented to our present environings but without an understanding of how the present orientation relates to our past orientation and environings. This capacity for *orientation systems-relations* has to be fleshed out.

Furthermore, without a *transcendent integrating system* we would have to rely on remembering a multitude of orientation systems and orientation systems-relations, which still does not explain how we might form a holistic understanding and which does not provide for the capacity for spatial improvisation. A transcending system must neutralize these differentiational qualities in an integrating way. The atlas metaphor is apropo to the sedimentation of these orientation systems in our memory. But the atlas would be more like a compilation of maze-like directives, if it were not for a transcending system-gestalt—the whole of which is different than the sum of its parts. This juncture seems to be the obvious source for the Cartesian xyz coordinate system as a thoroughly Objective system that absorbs the "infinity" of qualitatively organized coordinate systems. And it is this juncture that baited Husserl prematurely into the theoretical attitude. But as noted, an idealized mathematical system is an abstraction from experience and thus it is necessary to continue to consult experience rather than illegitimately allow mathematical theoretics to construct experience. First, I confront the problem of *orientation systems-relations*.

We cannot *merely* remember our past here, for the constitution of cognitive mapping cannot be achieved on the basis of temporality alone—in memory. Memory preserves an experience as past and the integrating relationship that is necessary is one constituted in a present moment. On the basis of my present here, I can presently coordinate my present zero point coordinate system with another that is a remembered-at-one-time-present-zero-point-orientation-system. But it is now interpolated into a new spatial context. This new context transposes its original zero coordinate point to a new spatial value on the basis of the present orientation system. However, the past zero point does carry a special qualitative value because it once was a zero point, a here. Let us call this remembered past coordinate zero point an *anchor point*. This is an apropo appelation because it remains in awareness as perhaps the *starting point* of the ambulatory project—*the journey*. At the past here I had set up my coordinate system and during the ambulatory project, I am aware of *heading out* in a direction *on the basis of* that orientation system. Through the *linearity of pathmaking* and out of the plethora of theres I am carving *directionality* in relation to the anchor, which eliminates the theres along the other possible directions. When I reach a there in which I want to take time to establish as a here, I know where I am on the basis of the past here as my anchor point along with its coordinate system. But something happens in the establishment of the new here; I make the new here the zero point of a new orientation system. Granted, it is the case that as I ambulate I am constantly converting heres and theres, but *ambulatory conversion* does not achieve the *connective function* of the anchor point. It is by constituting an anchor point that establishes a coordinate system by which I *qualitatively measure* my ambulatory project. In other words I treat my progressive heres while ambulating as if they were theres on the basis of my anchoring coordinate system. When I reach my destination and establish it as my new here, I set up a new zero point orientational system as my new anchor. My former anchor becomes a there in my present system, but retains its anchoring role: the-there-that-was-my-here. Other theres do not carry this quality of former zero coordinate point. This special quality of the *starting point anchor*, of carrying with it a different orientational system than the presently constituted system, overcomes the alienation of that system. I have my present orientation system based on my present here as my new *destination point anchor*, and my "other" present orientation system based on my past here as my *starting point anchor*. But the duality of systems still poses the same problem. How do both systems become integrated so that a higher achievement in cognitive mapping results?

The role of *circumrotation* must be examined. While circumrotating (in place) my here remains the same but the orienting coordinates exchange qualitative meanings through a continual axial transformation of right-front-left-behind, if I circumrotate in a clockwise fashion. The tree that was to my right is now in front of me, etc. In this case the zero point remains the same. So, how do I experience difference? I experience difference on the basis of the changing object-relations-to-my-Body. The *transmutation of directionality* is important for establishing the *directional modification* in ambulation. Because I can circumrotate and transmutate directions while maintaining my zero point, I am able to maintain the transmutational capacity in connection to that zero point *as my anchor* while I shift directions during my ambulatory project. As I turn left during my ambulation, I am able to envision that direction from my original anchor. Even though my Body is always here I think of my ambulating here as a there from the "standpoint" of my anchor, which is a past here. I place myself in its orientation system so that I can exchange my present ambulating here with it—the essential meaning of anchor has now been uncovered. An anchor point is an at–present-there of which I retain-as-my-here, and thus locate my present position from the standpoint of that anchor as the zero coordinate point. The capacity for the alienation of the Body's orientation system in the static here/there relationship allows for its own retrieval in the ambulatory project. The capacity for circumrotation is the necessary agency.

Two orienting systems are established on the basis of the two anchors, the anchor from which I commenced, *starting anchor point*, and the other anchor that functions as a *destination anchor point* (final or temporary). The former anchor point and the new anchor point that I presently establish at my destination allows for the *reversal of direction* in a new ambulatory project, the *return journey*, such that I can ambulate in the opposite direction back to the original anchor point. If I can do that with confidence, I am capable of imagining ambulatory directions in reverse ordering from my present anchor point, even with having no memory of perceptions walking in the reverse direction. What happens is that I can shift between the two anchor points as providing two *related* orientation systems and thus at any time I can locate my position from either anchor no matter what direction I am headed. The two systems and their *connection* have been established in my cognitive mapping. Even though the reversal can be imagined without perceptual experience of directional reversal, transcendence is not achieved. One merely recalls one's perceptions in reverse order and imagines on their basis.

Having established our anchor points does not limit us to our trail. When we started out from one anchor we were in the process of establishing a trail (path-making). Once we stopped and established another anchor the trail also was established. We may have been forging a trail but until it leads to "x" it had yet to become an *established trail*. "Where does that trail take you?" indicates this. Let us define an established trail as a set of ambulatory directions that have been established through traversal between two anchor points. Once the trail is established, then at any place along it we may at another time abandon the established trail and start off in another direction. The point at which we abandon the trail then becomes another starting point anchor, and an important one, a *nodal point anchor*, because we want to make sure we can get back, to connect with one or both of our *directionally reversible established orientation systems*. Nevertheless, we can still remember our original anchor point (original starting point) and its orientational system. Thus not only can we orient ourselves to the nodal anchor point we established when we left the trail, but we also can orient ourselves to the original anchor point. Moreover, we can also orient ourselves to the *original destinatory anchor point* that had orignally established our first trail and we can also orient ourselves to that. Thus, we have three *differentiated orientational systems* from which we can work. If we establish another destinatory anchor, which would establish the trail that we are now on, *an intersecting established trail*, we then will have four systems. But now it should be obvious that this juxtaposition of orientational systems cannot go on much longer. In other words one does not remain at this level of achievement whereby a catalog of orientational maps retained in memory is end of cognitive mapping. Rather an act that achieves transcendence and integration allows for the formation of one cognitive map so that we can be anywhere in the landscape and get to anywhere else without being confined to a set of anchor point orientation systems. It is then no longer necessary that we have already established a trail in experience, we can imagine a direction and we can imagine anchor points without a basis in past experience. On the basis of an integrated cognitive map, anchor points are assigned at will and the orientation systems are immediately projected onto it. But how is the transcendent and integrated cognitive map constituted? In what way do we maintain orientation while transcending the Body-landscape experiential basis and the Body's constitutive powers?

The bilateral asymmetry of the Body differentiates its morphological characteristics as front and back, due to the position of its sense organs, and its right and left, due mostly to handedness associated with work. As perceivable objects surround the Body, their whereabouts is assigned a direction—in front of me, behind me, to my right or to my left, and with all the gradations in between. These directions gather objects into my perceivable surroundings on the basis of a horizonal directionality. At some point the objects in these directions are no longer gatherable due to the limitations of the Body's sense organs, which means that *landscape continuation* transcends the Body's capacities for perception. This gathering around me in a zone of actual perception is recognized as limited, and my gaze turns outward into the horizon that opens to a transcending beyond. And, the embodied directionalities themselves continue into that horizon and beyond. Vision does not stop at objects in a particular direction, but rather intends the direction by which they are related to the Body. When perambulating the horizon moves also, which means that entities enter and pass out of the perceivable surroundings. They enter into view as they appear out of the horizon. Thus the directions that are based on the Body's characteristics always transcend the perceivable surroundings. So the ahead/behind and right/left that gather objects into the orientation system of the Body as the zero coordinate point become *outwardly bound* when gazing into the horizon. This outward directionality of the orientational system allows for the Body to constitute its own transcendence beyond its perceptual horizons, which bounds the immediate perceptual surroundings as actual and sets what is beyond as potential zones of perceptual experience.

Since the Body can circumrotate, these directionalities can be transmutated. Citing Husserl we have already discussed how the perceived objects form a system and that system corresponds with our kinesthetic experience such that a reliable fixed system of object-relations is formed. But does the transcending horizon move along with the transmutation of the directional qualities of the Body? We answer no, but how do we know? Certainly not on the basis of what is perceivable. If the answer is that through perambulation we find that what we have constituted as an anchor point remains fixed, this does not fix the horizon, for the horizon, again, always transcends us and so we never perceive its "other side."

The fact remains that we cannot constitute the fixed direction of the horizon on the basis of the Body alone. We need an Other. The Other this time is not supplied by another human being. The Other is the *Earthly-embodiment*. The EarthBody has its own characteristics, its own morphological differentiations and its own circumrotational capacities. As a celestial body it achieves its orientation from the arrangement that includes other celestial bodies. The Earth has a point of view, and orientation system, by which the other celestial body's can be oriented to it. Or, another celestial point of view can be taken and the Earth can be placed into its orientation system. A fundamental anchor for the earth is the sun, and the earth in turn serves as an anchor for its moon.

The Body on its own capacities can never settle the fixedness of the Earth's horizon. As a being of the earth, the Body's capacities depend on the co-constitutivity of the earth through the earth's morphology and its orientational system in order to develop a reliable cognitive map. My point is this: we are unable to create an integrated cognitive map that transcends the Body's multiple orientation maps without the co-constitutive agency of the EarthBody.

The EarthBody's fixedness is indicated to us by another dimension of human embodiment: above/below. The sky above, with the sun during the day and with other celestial bodies at night, indicates a fixedness that we depend on. As we establish anchor orientational systems horizontally across the earth's surface, we can establish their fixed positions in the transcendent horizon through directionality, which by taking up the *Earthly-embodiment's fixed orientational system*, we call the transcending directionalities north, south, east, and west.[24] But these transcending directionalities are not merely fixed through the EarthBody's morphology. Likewise, the Body's morphology alone does not fix left/right, ahead/behind, for these directions have no meaning unless related to perceivable objects in the surrounding spatial field, which are given as involving a lawful system of relations. For alone, these cardinal directions are thoroughly transmutable through our circumrotation. And, likewise the earth circumrotates. We do not experience the earth's circumrotation, but we experience what appears to the

Body's perceptual abilities, that is, the sun's direction across the sky. The EarthBody's orientation system in its relation to the sun constitutes for me the horizonal directions of east and west. East and west are transcending directionalities that unendingly penetrate the horizon. Heading in an easterly direction means that we are oriented toward the horizon in a certain way. This is the EarthBody's constitutive system for we are unable to perceive the transcending horizonal direction.

Through the EarthBody's orientation system in relation to our Body's orientational system, we achieve transcendence. We no longer need to remain wedded to the anchoring functions (starting, nodal, and destinational) that grounded our separate orientational systems. We now have something that fixes the horizon and transcends our anchors. For example, if we head out in a certain direction, we know we are heading easterly through the EarthBody's relation to the sun. The sun is a transcendent object, but perceptually present beyond my Body's perceptual relation to the landscape. By *heading* east we also become oriented to the horizons of west (behind), north (left), and south (right). But orienting to the position of the sun is to orient to a direction, for the sun cannot be reached; it never becomes a here. The sun functions as a *transcendent anchor*, from which emerges a *transcendent directionality.*[25] My Body does not constitute the directions, north, south, east, and west; these directions are constituted through the EarthBody's orientation system and its morphological characteristics.

The EarthBody's orientation system in relation to other celestial bodies allows us to transcend our own multifarious orientational systems in a way that we can integrate them. We achieve *Worldly-orientation*, which is the integrative *co-constitutive* relationship between the orientation systems of the Body in coordination with the EarthlyBody's orientation system.

From this anchor to that anchor, I set up two systems of orientation on the basis of these two anchors, starting and destinational, in their reversible roles. Now however I lift those anchors out of the Body's constitutive system by orienting them to the EarthBody's *horizonal constitutive system of the cardinal directions*. The anchors in such a system become related to a fixed transcendent directionality that *neutralizes* their orientational systems. Their function is no longer privileged, for they now are only two of many possible point locations. They become constituted in the *horizonal orientational system*, which means directionality is constituted for the anchors instead of the anchors constituting directionality. They are assigned positions along the cardinal grid. This means that the hereness of the Body that is the zero coordinate point of its orientation systems allows itself to be constituted by the EarthBody's transcendent horizonal orientation system. The resulting cognitive map is the co-participation of Body and EarthBody, which allows for the Body's anchor point orientation systems to be intgrated into one system. It is this cooperative participatory effort that "decenters" the Body and allows it to coordinate itself with the Earth's coordinates, to be a denizen at some specific place on the EarthBody from the standpoint of the EarthBody's morphology and coordinate system. That which is east is arranged according to the EarthBody's system and I superimpose that system on my anchor orienting systems so that they become determined on the basis of the earth's system. I arrange my Body's coordinates accordingly, allowing them to be constituted by the coordinates, that is, in a northerly direction or in a northeasterly direction. These directions are fixed, and the Body must orient itself accordingly and have the EarthBody's directionalities orient for the Body the possibilities for anchor points and trails. The two systems then integrate, and once the Body achieves cognition of the horizonal directionality of the EarthBody, all of its zero point orientation systems are integrated. If I intend to travel in an easterly direction my Body's anchor orientations are organized by the easterly system and they arrange themselves within that transcending horizon.

What allows for the Body's system and the EarthBody's system to become integrated is that the EarthBody's orientation system can be alienated. It is the sun functioning as the Earthbody's alienated zero coordinate that places the Body and the entities upon the EarthBody's surface (landscape) in an *integrating orientation system*. Both the Body and the entities it perceives are objects organized by the celestial body's transcending position overhead. It "looks down" on the EarthBody, whereas my Body "looks across" its surface. The Body takes the transcendent overhead point of view, which means where there is no anchor for it. From this transcendent position, "there," the Body envisions the EarthBody and its own location, from this transcendent above that is neither its own nor the EarthBody's, but from which it captures all of its own possible positions in relation to the morphology of the EarthBody.

The Earth's orientational system involves directional vectors with no zero point. Why? The EarthBody's morphology is spherical and as Aristotle has taught us we cannot find the starting point of a circle. The Body is no longer center in its system, but is to be directed out of itself into a horizon that it cannot experience. Nevertheless, the Body does not relinquish its own orientational form and in its participation with the EarthBody, learns the EarthBody's morphology by orienting that morphology to itself in various anchor orientational systems. But in turn the Body allows these orientation systems to be integrated by the transcending system of the EarthBody. Its morphology is fixed in its lawful relation to the transcendent horizon. Even though the sun rises in the east and sets in the west, the earth's morphology remains in a fixed order. When the EarthBody's system is alienated, we constitute the sun or some celestial body as the zero point of a coordinational field. In doing so, we can imagine an overhead orientatation, which is the experiential source for a representational map.

Thus, there is no intellectualist leap to a Kantian intellectual intuition and the Cartesian system of Objective space. The highest levels of cognitive mapping are achieved on the basis of the co-constitutive agency of the Body and the EarthBody in the co-participatory sensuous field.

Conclusion

In this chapter I have framed a phenomenology of cognitive mapping by beginning from a fundamental structure of qualitative spatial orientation, which has been accepted by many phenomenologists. I have shown that Husserl's treatment of this structure is inadequate for the phenemenological exploration of cognitive mapping. Due to a mistaken abstraction and a limitation within his methodological parameter (the Cartesian way or egological phenomenology) that Husserl employed at the time of exploring this fundamental structure of the zero point orientation system of the Body, I found it necessary to move beyond Husserl's discussion. I showed that the phenomenological framing of cognitive mapping has to include the EarthBody's orientation system as co-constitutive of the experience for the adequate description of the highest levels involved in the development of a cognitive map. This study is meant as a beginning exploration, which I hope will motivate further interest in the phenomenon.

Notes

1. I employ "EarthBody" as a technical term. I capitalize it to signify its constitutive function. I employ "Body" to refer to the human body's constitutive role and I use "body" to refer to the human body when it is meant as a constituted object. I would then employ "earth," if I were talking about the earth as a constituted object. The World in its embodiment is EarthBody.

2. See Edmund Husserl, *Ideas Pertaining to a Pure Phenomenology and to a Phenomenological Philosophy* Second Book, trans. R. Rojcewicz and A. Schuwer (Dordrecht: Kluwer Academic Publishers, 1989).

3. Alfred Schutz and Thomas Luckmann, *The Structures of the Life-World*, trans. Richard M. Zaner and H. Tristram Engelhardt, Jr. (Evanston Ill.: Northwestern University Press, 1973). Maurice Merleau-Ponty, *Phenomenology of Perception*, trans. Colin Smith (London: Routledge & Kegan Paul, 1962). Edward S. Casey, *Getting Back into Place: Toward a Renewed Understanding of the Place-World* (Bloomington: Indiana University Press, 1993).

4. Edmund Husserl, *Ideas Pertaining to a Pure Phenomenology and to a Phenomenological Philosophy* First Book, trans. F. Kersten (The Hague: Martinus Nijhoff Publishers, 1983).

5. There is far too much literature concerning cognitive maps and mapping to do anything but to provide a couple of examples of the sophistication in this study. Some researchers have developed the notion that the cognitive map is more than spatial acumen, it involves the spatialization of a belief system including meanings and values. See D. Wood and R. Beck, "Tour Personality: The Independence of Environmental Orientation in Interpersonal Behavior," in *Journal of Environmental Psychology* volume 10, number 3 (September 1990), 177-207. Also see Reginald G. Golledge and Robert J. Stimson, *Spatial Behavior: A Geographic Perspective* (New York: The Guilford Press, 1997). They write, "[Cognitive maps] contain spatial relational data along with environmental

attributes and individualized and socioculturally conditioned beliefs, values, and attitudes," 236.

6. T. Gärling, A. Böök, and E. Lindberg, "The Acquisition and Use of an Internal Representation of the Spatial Layout of the Environment During Locomotion," in *Man-Environment Systems* 9, 1979: 200.

7. Husserl, *Ideas II*, 61.

8. Husserl, *Ideas II*, 61.

9. Husserl, *Ideas II*, 61.

10. Husserl, *Ideas II*, 62.

11. Husserl, *Ideas II*, 63.

12. Merleau-Ponty makes much of this ambiguity of the lived-body—that it is neither consciousness nor thing. The ambiguity in its intentional capacity to constitute perceivable objects (Body-*Lieb*) and also to constutute itself as a perceivable object (body-*Körper*) distinguishes lived-body intentionality from that of consciousness. Consciousness can only constitute the object of its acts; it cannot intend in its own acts. See Maurice Merleau-Ponty, *Phenomenology of Perception*, trans. Colin Smith (London: Routledge & Kegan Paul, 1962). "When I say that an object is on a table, I always mentally put myself either in the table or in the object, and I apply to them a category which theoretically fits the relationship of my body to external objects," 101. I am labeling this capacity the alienation of the system of orientation.

13. Husserl, *Ideas II*, 166.

14. Husserl, *Ideas II*, 166.

15. Merleau-Ponty calls this the reversibility of the lived-body. Our one hand can either touch or be touched by our other hand. But the two hands cannot touch or be touched simultaneously. This structure of perception holds as well for vision. Either my own point of view is taken up or it is the point of view of the perceived object. It is impossible for both to occur simultaneously. See Maurice Merleau-Ponty, *The Visible and the Invisible: Followed by Working Notes*, trans. Alphonso Lingis (Evanston, Ill.: Northwestern University Press, 1968). "*Reversibility* is not an actual *identity* of the touching and the touched. It is their identity by principle (always abortive)," 272.

16. Husserl, *Ideas II*, 177.

17. Husserl, *Ideas II*, 177.

18. Husserl, *Ideas II*, 88.

19. Husserl, *Ideas II*, 88.

20. Husserl, *Ideas II*, 8.

21. Husserl, *Ideas II*, 195.

22. Husserl, *Ideas II*, 196.

23. Husserl, *Ideas II*, 195.

24. I am aware that different cultures may have other directionalities and that they may have more than four. The point here is that whatever the organizational schema, it involves an orientation that transcends the Body while linking to embodiment.

25. Again, I am using the sun for illustration. Other celestial bodies can serve the same purpose, which concerns the EarthBody's orientation system.

Chapter 5

The Die is Cast: Boundaries of Time, Boundaries of Space

John M. Rose

Boundaries are virtually always defined in terms of space. Indeed, we hold that space defines boundaries because boundaries appear to be spatial by nature. Yet is this indeed the case? Virtually all of our western thinking uses space as fundamental category and unit of analysis in constructing our understanding of everything around us. We think everything is in space, and we think in terms of space for everything. In opposition to the primacy and privilege of space in human understanding, phenomenology has elicited time as fundamental to human perception and knowing. Edmund Husserl, in particular, described both the temporal constitution of consciousness and the temporality of the objects in the world. Heidegger, as well, has described time as the condition under which entities presence themselves in being and under which we have an experience of those entities in their presence in the world. Despite the pivotal role of time exhibited by these phenomenologists, spatially oriented thinking still dominates the investigations and research of phenomenologists and the geographers, artists, historians, etc. whom they have inspired. Space continues to dominate our thinking about time as well. This article makes a few suggestions, through both examples and an analysis of time, for thinking in terms of time as an alternative for thinking in space. Specifically, this article frames an understanding of boundaries, something seemingly basic to space, in terms of time. Through exhibiting the temporal constitution of boundaries, I hope to show some of the aspects that thinking in terms of time reveals.

I will show that time is constitutive of boundaries in at least two ways. First, the temporal character of our consciousness is necessary to perceive or know a boundary at all. For example, one has to perceive the two sides of the boundaries to understand that it is a boundary, and this perception requires two temporally related moments in time. Second, the boundary itself also has a temporal constitution. The boundary is a real limit, and that limit has a temporal structure. The temporal constitution of the boundary as a real event (i.e., seeing the boundary itself in time instead of space and seeing how time constitutes the boundary in the world) is the more difficult aspect show. The temporal character of the boundary implies that it is a real limit, that is, the temporal boundary is something that cannot one cannot go past. The boundary is a limit situation.

These two aspects of the temporal character of boundaries seem to be in contradiction. How can the temporal boundary be a real limit situation, if one needs to perceive the two sides of the boundary to understand that it is a boundary? With the answer to this question we shall see the difference between the temporal constitution of boundaries and the spatial constitution. I will demonstrate that a fundamental difference between time and space is that, whereas there is always more space outside bounded space, there isn't more time outside bounded time. Rather, there is another, completely different time.

To exhibit the temporal nature of boundaries, I will give three examples: one from ancient religion, one from myth, and one from history. These examples will allow me to elicit how both the existence of boundaries and our understanding of boundaries as boundaries requires time. The examples will also provide the opportunity to show important historical, geographical, and cultural meanings that become clear when we explicitly put time at the forefront of our understanding. Following the examples, I will present a philosophical analysis of time that highlights the themes concretized in the examples. I develop Aristotle's analysis of time in such a way as to feature the characteristics that make up the temporal character of boundaries. Aristotle's analysis of time stands at the beginning of the long history in the philosophical understanding of time in terms of space. Aristotle's analysis is thus well suited for a reconceptualization of time without space as a unit of analysis. I will close with some reflections on the ways in which these reflections on time contribute to the enframing of a new dynamic geography of human events.

Introductory Considerations on the Privilege of Space Over Time

Conquering and managing space has been the ultimate goal of technologically oriented civilization throughout most of its recorded history. The land myths that lie deep in the memories of every European ethnic group belie a preoccupation with the autochthony of people to defined space. The imperial conquests produced by every example of European expansionist ambitions into Eurasia, Asia, Africa, and our own hemisphere, also evince the motivation of possessing delimited space. The instantaneous speed of technology that underlies the new conceptions of global economy, global civilization, global news, and global citizenship, which are now addressed in every undergraduate curriculum across the United States, is latest example of conquering space. Understanding space as something to be conquered and managed ultimately derives from tacitly holding space as an ultimate category of the real and thereby a fundamental unit of analysis. The west fundamentally thinks in terms of space, and consequently the west thinks of space as something that can be conquered, possessed, managed, and put to use for human activities. When we come to define the notion of "boundary," then, it is crucial to understand that we are tacitly employing a concept of space that tacitly manages competing human interests and purposes as defined by the appropriation and possession of space.[1] Yet the dominance of space in western thinking has yet to be evaluated thoroughly. Nor have we sought alternatives to the privilege of space in our thought.

One consequence of thinking exclusively in terms of space is the impoverishment of our capacity to understand time and to think in terms of time. The history of philosophy, for example, exhibits this impoverishment in almost every one of its episodes. From Aristotle's definition of space as a container[2] to Kant's "Transcendental Aesthetic,"[3] time is defined, explicated, and finally understood in terms of space. Space becomes the more fundamental concept, leaving us thinking of time in terms of some spatial configuration such as a line or circle. For example, time portrayed as a line has three features: it is one-dimensional; it extends to infinity; and it is continuous. Yet we overlook the one exception to the portrayal of time as a line. Whereas all the points on a line are present all at once, the parts of the time-line are present successively. As Heschel remarks, "The power we attain in the world of space terminates abruptly at the borderline of time."[4] Heschel's remark is very provocative for thinking about time boundaries. Time itself is a border that thinking in terms of space cannot transgress. What would such a temporal border, such a temporal boundary, be? What would it mean to think about, reflect upon, describe, and experience existence in terms of time and without the dominance of a thinking that is conducted with space as its ultimate category and unit of analysis?

I have chosen specifically to make a case for the temporal nature of particular kinds of boundaries in an effort to move towards a kind of thought that is explicitly structured more by time than space. We cannot easily give up thinking in terms of space, but we can think in terms of time as well. Time is present in a real way in reality, as well as in our thoughts. To make a case for this view, I focus in this article on the constitution in time of boundaries or limit situations. When temporal boundaries are transgressed, they exhibit the dual characteristics of irreversibility and irrevocability. In a sense that is not true of limits in space, boundaries in time are real limits that cannot be passed over or moved through. I need a temporally structured consciousness to conceive of this limit, but, whether I am conscious of it or not, time is a limit. The temporality of the world has moved on. This limit is most easily seen when one attempts a transgression of a temporal boundary.

The temporal character of boundaries will allow me to emphasize a new component of a dynamic human geography: time as a fundamental unit of geographical, historical, and philosophical analysis. My exposition will attempt to function at these three levels of geography, history, and philosophy. I will do by beginning with specific historical and contemporary examples of temporal boundaries. I then return to Aristotle's analysis of time to explain why these boundaries are temporal rather than spatial. With this excursion into Aristotle's philosophy, I then reorient thinking in terms of time rather than space. I conclude with suggestions about some of the features of a geography oriented around time that will be appealing to geographers, historians, and philosophers.

As it is now constituted, space virtually defines geographical analysis of boundaries exclusively and exhaustively. Boundaries are understood in terms of limitations in space. These limitations are defined as either natural, the limit of a natural phenomenon such as a river; or limenal, an arbitrary limit such as the Mason-Dixon Line. The identity of boundaries and limits is so determined by space that it is difficult to conceive of boundaries in any other way. Indeed, space itself seems to be ultimately determinative of human, physical existence.

One quality of boundaries, however, that does not seem boundary-like at all is that boundaries in space are so easily transgressed. We can always cross a river or drive over the Mason-Dixon Line. From this perspective all boundaries could be construed as limens or limits in a psychological sense. In what sense, then, are boundaries, whether natural or limenal, anything more than a certain kind of specification of space? For boundaries to have any kind of a more substantive meaning, there must be more to a boundary than space defined by a delimitation that can always be moved across. The problem here lies not in the reality of boundaries, but in the poverty of space as the dimension used to explain what a boundary is. That space is the sole dimension used to explain boundaries has more to do with the absolute character of space in our thinking about existence than it does with the nature of boundaries themselves.

Time affords a much richer way to think about boundaries. The transgression of a spatial boundary can always be reversed, but the transgression of a temporal boundary can never be reversed. Temporal boundaries are irrevocable and irreversible, while spatial boundaries are not. While these are features of temporal boundaries as such, they are not easily seen as indicative of the temporal aspect of boundaries until the boundary is transgressed. To illustrate the temporal nature of boundaries, I will now turn to three examples. I will describe each in turn.

Three Examples for Temporal Boundaries

My first example is the enactment of the Dionysian ritual in ancient Greece.[5] I will describe this enactment in such a way as to grasp the temporal character of the constitution of a boundary.

The dancers gather at sunset in a small hollow at the foot of a knoll. Several of the dancers drag a roughly oblong rock to the center of the hollow where they set it on end. Other dancers soon commence a rhythmic movement of their bodies around the stone. Simultaneously as a chorus, they start humming the dithyramb, the rhythmic, tuneless melody to which their bodies sway as they move across the sweet smelling grass. With their dancing and singing, the small hollow becomes a χόρα (chora), the dancing place. They successfully invoke the presence of Dionysus, and deer gather around them because of the presence of the god. The deer quietly graze around the dancers as the Dionysian ritual grows in intensity. Suddenly, the deer freeze in place, listening intently to sounds on the wind. Then they bolt away. The Dionysiacs know intruders are near at hand and run off. Dionysus returns to his cave.

Our word "chorus" evolved from the word χόρα, which refers to the scene of this collective enactment of the presence of the god. Χόρα in this context originally refers to the place of the dance, the dancers, and the dance itself. None of these three facets of the χόρα, however, is defined in advance. The dance was improvised and not orchestrated ahead of time. We should hear the root of χόρα in the word "orchestra," and we still preserve three meanings of chorus in the sense of "refrain," as well as the whole group which sings together and the part of the theater where the chorus once stood. Eventually Dionysian theaters were built for the ritual of enacting the presence of the twice-born god, Dionysus. In the early stages of the worship of Dionysus, the χόρα is defined by the physical movement of the dancers' bodies. The χόρα is constituted wherever the dancers dance. The movements of the dancers' bodies define the dancing floor.

The χόρα only exists while ritual is enacted. The χόρα only exists in time. Further, though elastic, the χόρα creates real boundaries that define not just a sacred space but a sacred time. The time of ritual is always sacred, and the Dionysian ritual is no exception to this rule. The time of the χόρα is the bridge between the mudane, temporal existence of the world and the divine movement of Dionysus' birth, death, and rebirth, which is the indestructible movement of life itself. This life the Greeks would call ζῷον, and it is different than βίος, life differentiated into different species or kinds of animals. The bridge that spans the gap between the mundane and the divine cannot be understood in terms of space. It is a boundary of time. The real presence of Dionysus, whose presence the Dionysiacs witness, is a temporal event.

The Dionysian ritual is a temporal event in two ways. First, it only exists for *a time* by creating a scared moment in time. This moment is a temporal boundary. When the interlopers transgress that time, the event is over; its time is over. The event cannot be retrieved or recovered. This irrevocable and irretrievable character is especially revealing when one reflects on the improvisational character of the whole event. We may suppose that no two Dionysian rituals are the same. Second, the ritual is a temporal event in that it transforms time from the mundane, worldly time of human existence with its birth and death and events as defined between into the divine time of the indestructible movement of life itself. Dionysus does not portend the individual's rebirth but the fact that life itself will bring forth a new birth of life out of death. These meanings of the

Dionysian are human truths that cannot be grasped by space. Time is the only dimension in which they cannot be fathomed.

My second example draws upon the temporal aspects in the story of the founding of Rome by Romulus and Remus. Again, I tell the story in the present tense and elicit the temporal character afterwards. Two brothers, twins in fact, decide to found a city. Both being fond of divination and sacred rights decide to look for omens and signs for their prospects for the city. Soon one brother sees six vultures, while the other brother sees twelve vultures. Though some say that the one brother did truly see six birds and the other feigned that he saw twelve so as to get the larger part, they determine from this oracle that each brother would be in charge of a section of the new city. The one brother, using a plow pulled by a white cow and a white bull, cut a furrow to mark the boundary of his section, but it was twice the size of his brother's section. Angry with his brother for getting the larger part, the second brother jumped over the furrow into his brother's section, where his brother then killed him. Thus, Romulus killed Remus at the moment of the founding of Rome. As for the day they began to build the city, it is universally agreed to have been the twenty-first of April, and that day the Romans annually keep holy, calling it their country's birthday.[6]

The significance of the death of Remus on the day of the founding of Rome lies in the temporal character of these acts rather than the space created by the landmark of the furrows. The story tells us that it is not enough to simply create a boundary in space. Innumerable furrows have been plowed without creating the boundaries of a city. The role of time in the story lies in the occasion of the dual acts of Remus leaping over the Romulus' furrow and Romulus killing Remus for the transgression. The time of founding is incorporated in these acts, which are measured again by time, in the duration of plowing for a day from the sun's rise to the sun's set. Parallel stories have other irrevocable acts, such as the women among the survivors of Troy burning the ships at Palatium because they were weary of travel. In them all, the importance of the occasion in time is marked by the character of an irrevocable act of leaping across the furrows and the murder itself on the occasion of plowing the furrows.

The third example comes from later in Roman history. It is the story of Julius Caesar's decision to become emperor of Rome. Because of crimes committed during a brief period as a senator, Caesar could not return to his native land without being prosecuted for those crimes.[7] Still eager for absolute power and influence, and wishing to become preeminent in the state that would prosecute him for his crimes against the republic, he obtains from his allies the post of governor of three provinces to the north of his state. Convinced that he possesses exceptional military and administrative talents, Caesar sets out to demonstrate his skill by first halting the westward migration towards the Atlantic coast of a people who had been driven out of their homes in Switzerland. Even though these people turned back and were no real threat to his order, this governor slaughtered tens of thousands of them. Over the next seven years of his administration, he crossed many geographical boundaries. He built a bridge over the Rhine river, setting foot where none of his country had ever set foot before. He twice crossed the English Channel, only to turn back after sudden storms devastated his navy. Yet even after seven years and many triumphs and fame, his enemies at home had not forgotten his crimes, and, were he to give up his command and become a regular citizen again, he would be tried and killed. His allies at home had also lost power and the ability to protect him, and the sensate passed an emergency decree with himself as its target. And so,

> On the night of January 10th, 49 BC, Julius Caesar crossed the small river Rubicon, which formed the border between Cisapline Gaul and eastern Italy. When he moved across the bridge, taking a single legion with him, he was breaking Rome's law of treason which forbade a governor to lead his troops outside his province—a law that Sulla had formulated precisely with the aim of curbing such acts against the government.[8]

Caesar's example is rich in its exhibition of the temporal nature of the boundaries transgressed. While the river is evident as a natural, geographic limit, its other obvious nature as water in constant motion provides us with a hermeneutic access to the temporal nature of Caesar's decision and the movement of determinative events in Caesar's life. As Heraclitus wrote: "You cannot step twice into the same river, for other waters and yet others go ever flowing on."[9] The events in Caesar's life are moving forward with such ineluctable swiftness that they are literally irreversible and irrevocable. Literally, he cannot turn back. Once Caesar has crossed the Rubicon, he cannot be other than what he now is: a traitor to Rome and its would-be dictator. Caesar enacts his own nature in his irrevocable decisions. The import of his act can only be seen in the temporal character of the boundary, and Caesar's life is a succession of such temporal acts.

These examples of the enactment of a sacred ritual, an act of violent passion, and a calculated decision initially present the creation of a bounded space and limit at a particular moment in time. Usually, the boundaries of such spaces for human events are treated as limenal and symbolic. They mark a temporary threshold between two spaces, a boundary that could just as well be elsewhere. Such boundaries are also interpreted as being highly permeable. Their permeability is understood in the sense in which these boundaries are also thresholds. They are boundaries that are meant to be crossed, allowing the movement through them of that which they only symbolically separate. The revelers dance creates a moveable, sacred space. Romulus' plow defines the furrow that Remus jumps across. The bridge is a passage over the Rubicon that Caesar must not take. From the perspective of space, the significance of the boundary lies in that it has been transgressed. The transgression, however, brings to the light the role of time.

As thresholds, however, these "landmarks" are no longer boundaries. They are marks of a passing time, one time into another. These three examples also present another characteristic that is not addressed in the limenality of these boundaries. Each of these boundaries and limits brings with it an implication that there is an aspect to the event that is irreversible and irrevocable. The sacred space of the Dionysian invocation cannot be profaned. Remus cannot be brought back to life. Caesar cannot be other than treasonous, or as Caesar himself put it, "*Alea iacta est.*" [The die is cast.] Each boundary has the character of a final decision about which the humans who occupy the defined space cannot reverse themselves. We cannot understand these boundaries and the spaces they define by interpreting them purely in terms of limenally defined space. They must be understood in terms of time.

Umberto Eco addresses the irreversibility of these limits in terms of the social contract. Eco writes,

> Even if these principles do not provide for the recognition of a physical order to the world, they do at least provide a social contract. Latin rationalism adopts the principles of Greek rationalism but transforms and enriches them in a legal and contractual sense. The legal standard is *modus*, but the *modus* is also the limit, the boundaries.
>
> The Latin obsession with spatial limits goes right back to the legend of the foundation of Rome: Romulus draws a boundary line and kills his brother for failing to respect it. If boundaries are not recognized, then there can be no *civitas*.[10]

Clearly, Eco is correct in pointing out that there can be no community if boundaries are not respected, but I do not think that Eco has gone far enough. These are not simply psychological boundaries or boundaries imposed by interpretation or social contract. The primary meaning of limenal is, indeed, the threshold of a physiological or psychological response. At least in their irreversibility and irrevocality, these boundaries are in some sense real, at least in the sense that they become something independent of what any of us might think of them. I think the shortcoming in Eco's remarks is that he continues to see these boundaries purely in terms of space, and, by seeing boundaries only in terms of space, we interpret them too statically. The reality of the boundary comes to light—as I believe it does in my three examples—when we are confronted with the transgression of the boundary, the unallowable movement across it. The reality of these boundaries cannot be grasped if we only interpret boundaries in terms of space.

By raising the issue of movement across boundaries, especially transgressive movement, we have opened the way to understanding boundaries in terms of time. Aristotle has already pointed out that time and movement are inseparable.[11] I will now develop Aristotle's analysis of time so as to show that the irreversibility and irrevocable character of boundaries can only be grasped by understanding boundaries as limits in time.

A Philosophical Theory of Time and Boundaries

My demonstration of the temporal character of these boundaries shows that these are boundaries in the sense that a limit has been reached in time, not space. As temporal limits, these boundaries are true boundaries in that they cannot be crossed. They can be transgressed, but that transgression only points out the temporal character of the boundary. The transgression points out where the limit in time actually is: e.g., in the moment of the irrevocable and irreversible act. It still remains to understand time in such a way that we can see both the temporal character of the limit that time creates and how there is a different time outside of that time. To find temporal limits, we need an analysis of time that shows the limit character in time. The character of time is thus the key in understanding the boundary nature of these events. For such an analysis of time, I now turn to Aristotle.

Aristotle's analysis of time, coming early in the history of philosophy as it does, has few philosophical antecedents.[12] Aristotle's analysis of time is relatively untainted by the preoccupation with space that characterizes the work of his successors on this topic. The slight tendency that we do find to interpret time in terms of space in the one reference to place, I will show, is actually at odds with the dynamic movement of time in the remainder of Aristotle's account. Yet commentators have characteristically seized on this reference to place to focus on space when reading Aristotle. I will use Aristotle's basic analysis of time to show the temporal nature of the boundary. As I go along, I will relate Aristotle's analysis to the examples above.

Aristotle's now classic definition of time is "the number of a motion with respect of the prior and posterior."[13] He derives this definition from the observation of any type of motion. He sees that any motion has a starting point and a terminus, which he terms the prior and posterior. The motion can be of any type from physical, external movement to the motion of internal mental or intellectual states. We should take the breadth of the types of motion Aristotle is willing to include literally. It is easy to lapse into interpreting Aristotle's account of time solely in terms of motion in space. Such a view would understand all motion as locomotion: that is, change of place such as linear or projectile motion. It is telling that Aristotle also includes the internal motion of mental or intellectual states as also being time. The intellectual motion of the soul or intellect is not simple linear motion in space; it is not change of place. Yet, such mental motion has a starting point and terminus, a prior and posterior, of a different kind. It has the starting point and terminus of the thought process involved in a decision. The character of a decision, such as Caesar's decisive act to cross the Rubicon or Romulus' act of murder, has the character of a motion of the soul. The enactment of a decision will help to show the nature of a temporal limit.

To get to this definition, Aristotle begins his discussion by analyzing time in terms of the starting point and a terminus of a motion. The starting point and the terminus are the moment when motion begins and the moment when motion ends, not simply the place where a motion begins and end. "Prior" and "posterior" in Greek have temporal denotation of time "before," "earlier," but also "further," "forward" and "latter," "last," "later," and also "coming after," "following," "after." Distinguishing this motion's "earlier" and "later" from a spatial understanding, we should say that the motion is enacted rather than placed. Motion is obviously dynamic rather than static.

The sense of "enacted" applies directly to my examples. The ritual that evokes the presence of Dionysus has a beginning and ending in this sense. The chora enacts the ritual as an event that begins and ends. Romulus and Caesar enact their decisions. The place of the ritual, the ill-fated leap, and the act of treason to the Roman republic, remain after the time of their enactment is over. Mere movement in space does not capture the sense of the boundary created by this enactment.

Having designated the interval of motion with "earlier" and "later," Aristotle states that the complete motion, of whatever kind, is a continuous magnitude. This continuous magnitude is equated with time.

> Since a thing in motion is moved from something to something else and every magnitude is continuous, a motion follows magnitude; for motion is continuous and because a magnitude is continuous, and time is continuous because a motion is continuous (for the time elapsed is always thought to be as much as the corresponding motion which took place.[14]

There are three terms of this analysis that are related to each other because they all show continuity: something moved, the magnitude of that motion, and the elapsed interval between the prior and the posterior. All three are analyzed in terms of being continuous, and Aristotle clearly thinks of them as going together in the same way. The motion is defined in terms of its start point and terminus, and thus magnitude of motion is defined by a prior and a posterior. The before and after of the motion are the moments that define the time.

Even though there is the strong temporal sense to "earlier" and "later" as defining the continuous magnitude of motion, Aristotle makes his one comment that links time to place:

> Now the prior or the posterior are attributes primarily of a place, and in virtue of a position.[15]

The reference to "place" refers to the definition he worked out earlier in Book Δ: Place is "the limit [which] coincides with that which is limited."[16] With this sense of "limit," we see Aristotle describing "place" as a container where motion takes place. The motion takes place in a container, and the container does not move: "Place is an immovable vessel."[17] Aristotle also writes: "A place, then, is this, namely, the primary motionless boundary of that which contains."[18] While place defines the motion, place itself is not dynamic. Place does not move. With the static character of place, we see the first hint of the spatialization of time.

I suggest that Aristotle himself thinks that place is too static for his analysis of time. Having mentioned the prior and the posterior as attributes of place, Aristotle immediately makes the distinction between the prior and posterior—or as I interpret them, the earlier and later stages—of the existing motion and the prior and posterior as essence or defining limit.

> Now the prior and the posteror exist in a motion whenever a motion exists, but the essence of each of them is distinct [from a motion] and is not a motion.[19]

As part of the actual motion, the earlier and later, the prior and posterior, are part of that motion. Yet this motion is not yet time. When the motion is perceived or understood to be defined "that is, when we perceive the prior and the posterior in a motion," then we have time "by specifying in it a prior and a posterior as its limits."[20] In other words, the prior and posterior are not distinct in the motion; they seem to be distinct in the perception of the motion. That conception is of the earlier and later as the limits of motion. Aristotle is worth quoting directly here:

> And we limit it [the time] by believing that they are two distinct things and that there is something else between them; for when we conceive the extremes as being distinct from what is intermediate and the souls says that there are two moments, one prior and the other posterior, it is then that we also say that this is time, for what is limited by moments is thought to be time, and let this be assumed.[21]

The earlier and the later have to be conceived as distinct aspects and limits of the motion. The earlier and later are thus the boundaries and limits of the motion.

The earlier and later of the motion are clearly not spatial boundaries, for the motion has begun and ceased in moments not in space but as time. We can see that this is for several reasons. First, the "earlier" and the "later" are only distinct and separable in the conception of the motion and not the motion itself. Second, since Aristotle also thinks of intellectual motion or motion in the soul as subject to the same analysis, it is not likely that "earlier" and "later" are to be construed as places. If they are at all like place it is only in the sense that they are taken to contain the motion in the conception of time, and not in the actual motion. "Earlier" and "later," "prior" and "posterior," are temporal limits. It is important that in the actual existence of the motion "earlier" and "later" are not separable moments; they are part of the motion. It is in perception, in a conscious act of consciousness, that they become the limits of the motion. This is our first hint of a theory of the temporal character of boundaries: boundaries are the perception of a limit as an earlier and later of a magnitude of motion. Boundaries in our experience are moments.

The numbering of the "earlier" and the "later" is the last phase in Aristotle's analysis. Aristotle specifies two senses of "number": that which is numbered or numerable and that by which we number.[22] "That by which we number" is of course the numbers themselves or quantifiable units that we use to number.

"That which is numbered" is something that gets numbered or enumerated. The earlier and later moments, the starting point and terminus are thus numbered. Again, the actual motion, whether physical or intellectual, is continuous; the earlier and later moments are perceived as being the limits of the motion. What is numbered are the earlier and later, or prior and posterior, perceived as the limits of the motion:

> So when we perceive the moment as being one, and not as prior and again as posterior in a motion, or as being the same thing, but as prior to something and as posterior to something else, then no time is thought to have elapsed since neither was there a motion. But when we perceive it now as prior and later as posterior, then we say that time has elapsed. For time is just this: The number of a motion with respect to the prior and the posterior.[23]

Aristotle distinguishes the perception of the moment that limits two distinct moments from one and the perception of a moment as the beginning of one continuous movement. No time needs to elapse between two motions; they could indeed be concurrent. It is when the prior is perceived in relation to the later as one continuous and whole motion that the enumeration takes place. For example, we drop a ball from our hand and count "one" when the motion starts and "two" when the motion terminates. The enumeration of the prior and the posterior as boundaries defines the magnitude between the boundaries as elapsed time.

That the earlier and later are numbered implies the reality of the earlier and later moments as temporal limits in the motion. If the motion is ongoing in the sense that it has not yet reached its terminus or limit, or if there is no motion at all, there is in Aristotle's analysis no time. Again, if the earlier and the later moments were not real, that is, if they were not there to be enumerated, there would be no time.

To illustrate this point, I suggest my reader experience a muscian counting off the tempo of a song for a musical ensemble. She would begin by humming the tune out loud at the tempo she would like to set. The humming of the melody to herself would be the required motion. Then, while humming the melody again, the muscian would count off the tempo for the band. The counting of the tempo is the enumeration of the temporal limits, the temporal perameters of the time of the song. The counting off of the time of the song does not create limits. Those limits are already present in the tune itself. The counting enumerates the limits, showing them to the other members of the ensemble. The time of the song are limits that the band members will mutually function within in the performance of the song. These limits are not only in perception or conception; they actually exist in the tempo of the song.

Aristotle's analysis of time thus points out features of a continuous magnitude of motion. In relation to the magnitude of motion, Aristotle highlights the perception of two reference points that limit the magnitude of motion. This perception has three components: 1) the perception of a motion as a continuous whole;

2) the perception of the magnitude as limited by specifying the prior and the posterior as two distinct moments that the motion spans, and 3) the enumeration of the prior and the posterior.

Aristotle explains this perception by working out the analogy between motion and magnitude. In doing so, he explains the analogy between spatial boundaries of motion and the enumeration of temporal boundaries of magnitude.

> So in these, too, when existing as a being, the moment is the same (for what is prior or posterior is the [same] body in motion), but in essence it is distinct, for it is qua numerable that the moment is prior and posterior. And it is this that is most knowable, for a motion too is known because of the body in motion.[24]

As boundaries of the motion, the prior and the posterior are part of the actually existing motion, for motion is continuous. The perception, however, recognizes the boundaries of the magnitude as two distinct parts with the continuous time between them. In perceiving time, the perception thus changes the way it knows the starting point and terminus of the motion. The perceptive act sees the prior and the posterior as two separate limits of the magnitude instead of being continuous with the motion.

The upshot of this for our purposes is that the perceptive act knows time in a different way than it knows space. We need three points to define space, but only two—a prior and a posterior—to define time. The continuous nature of the motion is already limited by the beginning and end of the motion. The perceptive act specifies and numerates these extremes as limits that are really there. The perceptive act enumerates the prior and the posterior as boundaries, thus completing the cognizance of time. The act of enumeration specifies the whole magnitude in terms of a discrete interval, delineating the boundaries as the temporal limits of the magnitude. The prior and the posterior are thus two different moments that define the magnitude, but continuous in terms of the motion. As different concrete motions are defined by different starting points and termini, these different motions are literally that: different motions, different magnitudes of motion, different limits, and hence different times.

In the examples given above, the events are set off and limited by this temporal constitution of the limit to their motion. In our perception we see bring out the temporal limits of the events. The Dionysian ritual has its prior and posterior in the commencement of the ritual and the intrusion of the interlopers. The god is present to the Dionysiacs within the temporal limits of the ritual and not before or after. Remus' act of leaping across Romulus' furrow is bounded by the temporal limits of the beginning and end of the leap, and Romulus' act of murder is a separate time bound by different limits. When Caesar crosses the Rubicon with one legion he leaves the province of which he is the lawful governor and enters eastern Italy as one who has committed an act of treason. These movements are defined by their temporal limits, and these temporal limits are there to be perceived and enumerated.

My version of Aristotle's account of the defined temporal magnitude also lends itself to addressing the standpoint on the enumerated limits of these motions. Time is different in terms of each motion and each determined magnitude. Thus, the standpoint from which the determining enumeration is accomplished is not permanently fixed in a specific location of a static perceiver on the motion being defined. Like real human beings, the perceiver is able to move. The standpoint of the observer comes into being with each act of enumeration. The standpoint on temporal limit comes into existence with the enumeration of the limited magnitude. In a word, the standpoint is "evanescent." The magnitude is defined by both the prior and the posterior and the enumeration of these as limits. Rather than time being a creation interior to the intellect, creating an ideal concept of limit with which we label the motion, this perception of the temporal limit brings the standpoint of the observer into being. The temporal limits are already there, it is the perception of those temporal limits that establish an evanescent reference point by enumerating the limits of the magnitude of motion.

This act of enumeration is not the same as holding up events against a yardstick in a unitary and coherent time-system. Aristotle, himself, is remarkably unconcerned with a concept of time that serves as a measure in this way.[25] Still, this act of enumeration has universal applicability as a definition, without the claim that the time measured is a universally coherent, integrating time-system. Such a universal time would, on Aristotle's account, require an ultimate and absolute prior and posterior, which on his claim of the eternity of motion implies that these do not exist.

In discussing what he calls "the retrenchable now," G. E. L. Owen has also noticed something like this act of enumeration in Aristotle's account,[26] but Owen still argues for the parallel between temporal magnitude and spatial extension. The "retrenchable now" is Owen's term for the different magnitudes determined by different evanescent reference points. He notes that the temporal magnitude is not further divided into increments or units that are the mark of the time-line: "Aristotle's argument leaves us with no lines in time present, and only present moments? And all we can know is that one is earlier than another."[27] Owen suggests that we interpret Aristotle's moment or now as the counterpart to the spatial here. The spatial 'here' Owen understands as pointing out any stretch of space which can be narrowed or increased depending on what one wanted to point out. The magnitude of space would be defined by whatever boundary, or to use Owen's term, "frontier" one chose. Owen explains,

> There is no one reason for setting the frontier closer or farther: the demonstrative carries no privileged reference to a measure or ground that could not for another purpose be narrowed. Let us say, for short, that "here" is retrenchable.[28]

The temporal *now*, like the spatial *here*, is a magnitude defined by boundaries. Owen points out that these are temporary, but not temporal, boundaries that one employs for the purpose of reference, as we would when we say, "My property is here." Further for Owen, the sense in which we use the terms, *boundary* or *frontier* is not objective. When applied to the temporal now, what a speaker counts as present would be defined in terms of what he/she counts as past or future or both. "To understand 'here,' as to understand 'now,' the idea of retrenchability for different purposes is essential, and that of picking out an unextended point is not."[29] Owen, however, says that this comparison is not exact: space remains accessible in a way that time does not. Unlike space, what can be counted on any occasion as time past is irretrievably past. Thus, past and future are not strictly comparable to left, right, and yonder.

In this last point, however, Owen has missed exactly the temporal aspect of boundaries that I am describing by means of Aristotle's analysis. Owen argues that we can move the now, even though time is different from space in that space outside of the defined space remains accessible to perception. First, we should not overlook the dissimilarity between the temporal *now* and the spatial *here*. Boundaries in time are fixed in the act of perception in terms of a specific magnitude of the motion, whether that motion is external and physical or internal and mental. The prior and the posterior—without the time-line—have no external coordinates in terms of which they can be arbitrarily moved. Thus, what lies outside the temporal coordinates remains indeterminate and is not time, while what lies outside the spatial frontiers is still space. This atemporal indeterminacy can become time in relation to another magnitude, but that requires another act of perception and another reference point.

Second, Owens has clearly fallen into spatially oriented thinking. The accessibility of space for narrowing or expanding boundaries means that everything in space is relatable or mappable in terms of coordinates drawn by lines. Time as succession is converted into space when it is coordinated to the time-line in a unified time-system. The evanescent reference point is indicative of the way time is not like space. Not everything in time is directly relatable in terms of some permanent larger time. No time is any other time; it is not any other time except itself.

Boundaries in Time

Thus far I have presented both examples of temporal boundaries and a theory of time to show how boundaries are real limits constituted by time in our experience. Boundaries have a temporal nature that is overlooked by thinking too much in terms of space. I will now conclude by applying my theory of time in such a way as to illustrate the temporal character of boundaries further and extend my examples. The key to this application lies in pointing out in concrete illustrations of the act that perceives the temporal magnitude in the continuous motion. In other words, I will apply the act of perception that discloses the magnitude to an actual motion across the spatial boundary so as to see better the boundary in terms of time. I have selected two further concrete human actions or performances. Taking my previous example of the leader of an ensemble counting off the beat as a starting point, I now suggest that we understand the perceptual act that perceives time as real boundary in terms of a performance or enactment. The performance or enactment shows the perceptual act that discerns the temporal boundary in an embodied way. The enactments I have selected that embody that act of perception are the decision and the ritual.

The person making a decision perceives and responds to boundaries in terms of time by operating within a continuous, finite movement of the flow of clock time or calendar time. This operation involves the perceptual recognition of a temporal limit in precisely the way that I have described the act of perception of a temporal magnitude, that is, a perception of a real boundary in time. To be very precise, I am referring to the actual act of making a decision, not the situation of being compelled to decide.

Before the occasion of a decision, there are no temporal boundaries to be discerned. The motion of events is only potentially marked by as yet unspecified, or not-yet-perceived, limit-points that will be the determinations of the temporal boundary. I can choose any of the mundane and routine limits that present themselves in perceptions of time: before and after noon, night, sleep, breakfast, the finishing of the project I am working on, etc. The possibilities are unlimited for the perception of limits of temporal boundaries, which is perhaps why we so rarely think of boundaries in terms of time. Upon the occasion of making a decision, the unlimited possibilities for the determination of temporal boundaries become concrete, determined, and specific actualities in the moment of decision. The decision is the recognition of and acknowledgement that the limits of time are there. We often speak of the time of decision, meaning that limits have been reached that set this time apart. We cannot go back and change what has been decided, changing our mind is effectively making a new decision.

In addition to being an act, the decision is the acknowledgement of a temporal demarcation. The decision creates a boundary in which everything is determined in the mode of the decision, and it is an irreversible and irrevocable boundary and limit that we cannot cross. Caesar cannot go back over the bridge the next day and say he did not mean it. Remus cannot jump back over the furrow. The Dionysian revelers cannot turn their gathering into a Tupperware party. The movement of the body embodied in a decision creates a boundary that is not adequately described by the movement across space.

We also see the act of perception of a temporal boundary in the enactment of ritual. Ritual, especially, is the enactment of time in space. In ritual, we thus see a space created and endowed with the characteristics usually associated with boundaries, but space is created solely through the recognition of temporal boundaries. The space of ritual exists only for the time of the ritual enactment. In the call to the prayer, in the dimming of the lights for the theatrical production, in the religious ceremony, the space is transformed into the space of ritual through the initiation of the moments of beginning and ending of the ritual. These moments are the delimitation of the temporal magnitude of the time of the ritual. Hence, the temporal character of the boundary is even more radically illustrated than in the example of the decision. Ritual time defines the space that separates the scared from the profane through time. It shows us as well the role of the significant, ritualistic act that creates the time that separates the irreversible ritual from the mundane and reversible activities that can be taken back. When the ritual is over, the boundary created by the ritual no longer exists. The fourth wall in theater, for example, the wall through which the audience views the production, only exists while the performance is in progress. The synonyms I have been employing for performance, such as *production* and *enactment*, testify to the creation in time of the space that is bounded by a temporal act.

Evidence as I have tried to provide here, as well as theory to support that evidence, of the temporal character of boundaries can still easily be taken as evidence for the view that the constitution of time is a mental or psychological act that is interior to human consciousness and not indicative of time as real limits in our experience. Indeed, as I argued at the opening of this paper, space is so fundamental a term in our thinking that it is hard to see our situations defined otherwise than in terms of space. The recognition of the temporal character of boundaries can indeed look very much like an act of creating the boundary through an act of mind or will. To reorient our thinking it is key to keep in mind that, while space is always present to our perception at once and as a while, the whole of time is never present to us. We only perceive time in moments, in spans, in magnitudes defined by limits. The force of these limits is more consistent with the nature of boundaries as limits then are the highly permeable and limenal boundaries of space. The act of recognition of the temporal limit is seated in a human act of consciousness, and these temporal boundaries that are so recognized are real. Though time has too long been understood in terms space from Aristotle's analysis until Kant's first critique, the question I suggest we should pursue is how human consciousness remains cognizant of time while it continues to rely so heavily on space to represent the world. The limit events in human life, the events of birth, death, and the momentous decisions in a life are the true boundaries that cannot be surpassed. We celebrate or fear such events precisely because they are real limits that cannot be transgressed.

Notes

1. Consistent with my claims, I also note that the Supreme Court's decision on abortion rights in Roe v. Wade was determined according to the concept of tenancy and the right to private property. That is, the woman's relation to her body, specifically her womb, is that of the owner of property, viz., privately defined space. The fetus' relationship to that property is that of a tenant who has no rights to that defined space.

2. Aristotle, *Physics*, trans. Hippocrates G. Apostle, (Grinnell, Iowa: Peripatetic Press, 1980), Book Δ, chapter 5, 212a 30-213a 12, 68-69.

3. Immanuel Kant, *Critique of Pure Reason*, trans. Norman Kemp Smith (New York: St. Martin's Press), 65-81.

4. Abraham Joshua Heschel, *The Sabbath* (New York: Noonday Press, 1975), 3.

5. Carl Kerényi, *Dionysos: Archetypal Image of Indestructible Life*, trans. Ralph Manheim (Princeton: Princeton University Press, 1976). Indra Kagis McEwen, *Socrates' Ancestor: An Essay on Architectural Beginnings* (Cambridge, Mass.: MIT Press, 1993).

6. Plutarch, *Lives, Vol. I*, trans. Bernadotte Perrin (Cambridge, Mass.: Harvard University Press, 1914, reprinted 1998), 115-119.

7. I take my account from Michael Grant, *History of Rome* (New York: Charles Scriber's Sons, 1978), 216-217, 227.

8. Grant, *History of Rome*, p. 227.

9. Heraclitus, frag. 21. See also, frag. 24. "Time is a child moving counters in a game; the royal power is a child's." (52)

10. Umberto Eco, "Interpretation and History," in *Interpretation and Overinterpretation*, ed. Stefan Collini (New York: Cambridge University Press, 1992), 27.

11. Aristotle, *Physics*, "Now together with a motion we sense time also. For even when it is dark and we are not being affected through the body but some motion exists in the soul. We think without hesitation that along with that motion also time has elapsed; and further, when some thought is said to have elapsed, it appears that some motion has occurred simultaneously. Thus, time is either a motion or something that belonging to motion; and since it is not a motion, it must be something belonging to motion." 219a 4 -10.

12. Aristotle mentions these two previous accounts of time in his customary survey of his predecessors.

13. Aristotle, *Physics*, 81, 219b 4.

14. Aristotle, *Physics*, 80, 219a 11-15.

15. Aristotle, *Physics*, 80, 219a 16-17.

16. Aristotle, *Physics*, 212a 30.

17. Aristotle, *Physics*, 212a 15.

18. Aristotle, *Physics*, 212a 20.

19. Aristotle, *Physics*, 219a 20-22.

20. Aristotle, *Physics*, 219a 22-24.

21. Aristotle, *Physics*, 219a 25-30.

22. Aristotle, *Physics*, 219b 7-8.

23. Aristotle, *Physics*, 219a 30-219b 3.

24. Aristotle, *Physics*, 219b 26-31.

25. Cf. Louise Robinson Heath, *The Concept of Time* (Chicago: University of Chicago Press, 1936), 79, "The problem of measurement seems to have been formulated less clearly by the Greeks than any of the other problems relating to time, unless it be that of subjectivity and objectivity." And, "the neglect of measurement may be due to the Greek failure to draw clear distinction between the subjective and the objective."

26. G. E. L. Owen, "Aristotle on Time," in *Articles on Aristotle*, Vol. 3 "Metaphysics," ed. J. Barnes (London: Duckworth, 1979), 149-150.

27. Owen, 150.

28. Owen, 149.

29. Owen, 149.

Chapter 6

The Geography of Material Artifacts and an Outline for Synergetic Geography

François-Xavier Nzi iyo nsenga and Gary Backhaus

Introduction

This chapter construes a new classificatory approach to the geographical sciences, which is framed as a broad, suggestive sketch. The first section argues for the delineation of the Geography of Material Artifacts in distinction from Physical/Natural Geography and Human Geography. This proposal for a threefold classification is based on the demarcation of ontological regions. We define Ontological Geographies as dealing with perceivable and indeed mappable entities along the earth's surface. An essential principle delineating the ontological region of material artifacts is offered. The second section of the paper argues for a Synergetic Geographical Science, which functions to provide a common language making it possible to establish the interrelations between the ontological geographies. But it also functions to apprehend deeper levels of realities that are the genetic source for the manifestation of the perceivable entities comprising the Ontological Geographies. Synergetics involves a transdisciplinary approach to the emergent processes and structures of self-organizing and open dynamic systems. Particularly useful, Richard Buckminster Fuller has created a language of synergetics that best supports our view of the need for a well-developed Ontological Geography of Material Artifacts. Fuller's synergetics revolves around local technologies, or what we could call, "cosmological makings," which are the dynamic manifestation of interpatterning complexes of energy at all systems-levels of the Universe. This language allows us to link the ontological geographies to the idea of a Synergetic Geography. Synergetic Geography deals with the topology of synergetic systems and the mappings of their energy flows. Synergetic structures, which organize the flow of energy resulting in entity perdurance and decay and the ecological relations between entities, interrelate the Ontological Geographies on a common basis and complement them by capturing organizational complexes that transcend the limitations of human perception upon which the ontological geographies are based. The third section considers the scope of the Geography of Material Artifacts in relation to the other ontological geographies and to Synergetic Geography.

The use of material artifacts is necessary to human life, which is accomplished by appropriation, apportionment, containment, and exchange of cosmic energy circumscribing the earth's landscapes and atmosphere. The "spacings" of human beings, or any beings for that matter, concern more than "thingly" boundaries; it concerns the dynamic interrelationships that are formed with the environs on which individualized entities depend for their relatively stable perdurance. Material artifacts have in all times constituted the prime aid in the physical and psychological well being of the human species. Even prior to the spoken word, they were the structural backbone for the constitution of human societies and for individual survival. Humanity has transformed much of the earth through the objectivation of its meaningful expressivities into material artifacts. Human transformation has become so voluminous and intensified that its effects threaten the planet with grave and destructive consequences. Thus, much greater care, concern, and responsibility must become instilled in humanity's valuings in terms of its artifactual productions. We argue that it is necessary to create a geographical science of material artifacts, because it is vital that we no longer continue to subordinate our artifactual transformation of the earth under existing rubrics. With the advent of the consumer society and the increased mobility of humans, material production has resulted in landscapes being literally littered with quantities of material artifacts in a way never before experienced in human history. So far this new phenomenon has mostly captured the interest of ecological activists. However, ecological attitudes must become a component of all sciences, and the mapping function of geography is of utmost importance to even grasp the enormity of our material productions.

Only a short time ago, the spread of material artifacts was limited in scope, both physically and conceptually, and it was of a relatively slower pace. Since the industrial revolution, however, with the increase of material goods, mankind happens to be more and more privileging and clinging only to limited forms of otherwise abundant mental and physical structures. We are thus depriving ourselves of a varied array of alternatives, which might have been our best guarantee for survival and enrichment in a high quality, ever-versatile ecosystem. There is, in the view of many, a true and increasing threat in the present exponential accumulation of artifacts on earth. Mankind may soon suffocate under environmental pressures exercised by the increasing quantities and poorly conceived qualities of material artifacts. The cruel irony of the consumer mythology is the suicidal affluence that so far characterizes the twenty-first-century "global village." Mankind continues hoaxing itself in a gravely detrimental illusion of limitless environments, both conceptually and materially, which is leading to the worldwide indulgence in destructive excessiveness.[1] Energy containment for ecological health is overlooked; the trend is still the continuation of growth with no limits! The resulting effect is that environments are more and more replete with quantities of no longer used, little used, and irresponsibly over-used material artifacts, directly or indirectly detrimental to ecological and moral integrity.

How can the continuing ecological catastrophes be prevented? Geographers strive to analytically describe, as precisely as possible, each of the territorial features, which Jacques Berque broadly summarizes as land, habitat, and bordering areas.[2] The analytical approach alone is insufficient in terms of its descriptive objectives when relatively large-scale interrelated phenomena are considered. The analytical approach is found inadequate in providing sound explanation, let aside prediction, of how to approach maintaining, re-establishing, or making healthy ecosystems. In agreement with Jacques Berque and others advocating transdisciplinarity[3] what is now missing is the "commutation" approach throughout all the spatial features, leading to a fuller comprehension and, correspondingly, to more appropriate and efficient interventions into the ecological processes. Thus, with this diagnosis of the ecological crisis as the consequence of material artifactual excess, we advocate the necessity of delineating the Geography of Material Artifacts as a fundamental geographical science, which should take its place as a separate discipline alongside Physical/Natural Geography and Human Geography. We argue that this classification is an analytic and ontological necessity. Ecological interrelations cannot be properly framed unless the "elements" of the interrelation are recognized. Once analytical demarcations of ontological regions are better conceptualized, "commutation" takes into better account the ecological processes indicated by the ontological manifestations.

The Geography of Material Artifacts

In his foundational publication, Richard Hartshorne has defined the function of geographical science as follows:

> Geography seeks to acquire a complete knowledge of the areal differentiation of the world, and therefore discriminates among the phenomena that vary in different parts of the world only in terms of their geographic significance *i.e.*, their relation to the total differentiation of areas.[4]

Significant differentiations in the features of the earth's areas are of several kinds and they could be integrated in a plethora of ways. While some scholars have been concentrating on the study of purely physical features, both organic and inorganic, and differentiating geographical regions, others have focused on specific human social and cultural features in regards to those landscape characterizations. Population size and spread, development and change of cultural elements, political organizations and control, and economic exploitation and distribution are thus studied as some of the features in human geographies.

In the physical and human geography disciplines, the following three questions are considered to be basic: Where is it? Where are things located on the surface of the earth? What is it like There? What are the characteristics of a location or region? Why is it There? Why are those things located there?[5] The emphasis in classic geographic studies has been on characterizing *where* and on *there* of the triptych. As for the *it* of the above triptych, in other words, the precise nature of the locators, it has been left mostly to other disciplines—geology, biology, sociology, anthropology—to all the relevant natural and human sciences. Geographers have depended on the findings provided by those others disciplines, with little interpretation of those findings in relation to the intrinsic characterization of earth locators. Enough attention has not been given to distinguishing an ontology of material artifacts, also occupants in various earth locations. This has led geographers to overlook material artifacts, subsume them when appropriate under physical geography, and more frequently subsume them as subordinate denizens of human geography. The idea that material artifacts constitute an ontological region has not been developed in theories of geography and is underdeveloped in the philosophy of science. It seems that studies have not been undertaken to elucidate the specific nature of man-made objects, material artifacts, in their quality as independent locators on the earth's surface, that is, as a demarcated ontological category with *significant areal differentiation features* worthy of special study by geographers.

Certainly ecology has provided great service in not only treating the relations among living entities, but also in uniting various sciences so that their treatments are less abstracted from the concrete whole. Similarly, we believe the Geography of Material Artifacts not only plays the role of delineating an ontological region, but also functions to determine a whole to which various forms of scientific investigations relate.

Since an ontological region of material artifacts has not been propounded, studies of material artifacts are scattered amongst the various disciplines. Material artifacts have been mainly studied through physics, chemistry, and mechanics. Ergonomics endeavors to fit material artifacts to those who use them in their daily livelihoods. Material artifacts are also at the center of studies in statistics of distribution and purchase of man-made goods. In these statistical approaches, the main interest is more in accounting the debit and credit of merchandise and monetary costs, and not in characterizing the materiality of artifacts. Material artifacts are also secondarily considered in ethnology, anthropology, and semiotics. Traditionally, in those social disciplines, the concern has mostly been the human being, with his implements studied tangentially as obliged appendices.[6]

Various cultural interpretations can be given for this oversight. Historically, Western culture has been predominantly *androcratic* with a hunter-purveyor's sensibility. Material artifacts have been considered as the expression of physical, psychological, and socio-political power, yet less prominent than organizational strategies and acquisition techniques. Whereas the *gynecoid* life-bearing, protective, and reproductive sensibility along with domestic implements of daily use are subordinated to the showy outward paraphernalia of the explorer, the military-administrator the trader, and the "conspicuous consumer." They are personal, and their productiveness is concealed primarily for safety and survival reasons. Along with women, children, and slaves, material artifacts are the male's private property and therefore not subjected to public exposure except in particular moments of their production and acquisition, and eventually in trade, legacy, and parade. Another perspective suggests that daily life material artifacts historically have been assimilated to the vicarious physical work of slaves and *inferior* social classes. The *noble* intellectual work of acquiring and organizing territorial occupation and artifacts has always been the sole self-bestowed exclusivity of *superior* (meaning masculine and powerful) social classes. As a result, a haughty attitude towards ordinary material artifacts did not foster interest in their study, while the other products of the human intellect, the non-material or metaphysical artifacts, were intensively studied. A final explanation for the non-treatment of material artifacts is the idea that the ontology of material artifacts would lead to an anthropomorphic emphasis on materiality. It is feared that such an approach to material artifacts would further the abhorred yet rampant fetishism.[7] Recently though, Risto Hilpinen has proposed to conceive of them as *inhabitants* of the spatio-temporal world.[8]

We hereby propose to amplify this view and submit the following thesis for philosophical and geographical inquiry. Besides being mere aids to human psyche and to human bio-physical and social agency as alluded to above, material artifacts are in their own right also essentially important locators of the earth and its atmospheric layer. No matter how small or large, they all occupy portions of the total cosmic area. A thumb pin and a microchip occupy certain proportioned spaces, as do roads, bridges, buildings, rivers and mountains, fauna and flora, and human beings. Consequently, in that common quality as spatial entities, but essentially differentiated as objectivated products of human techne, material artifacts also can and should be referred to as "significant elements" in geographic studies.

Since the ontology of material artifacts has not been developed that would lead to the Geography of Material Artifacts, many studies focus on the expressive process in the objectivation of human intentionality. Thus, material artifacts are treated in a way that subsumes them under Human Geography. Randall R. Dippert defines artifacts in a trichotomy of *Instruments*, as Things Intentionally Used; *Tools*, as Intentionally Modified Instruments; and *Artifacts Proper*, as Objects Intended to Be Recognized as Tools.[9] His emphasis is on man-made material entities only from the perspective of the human agents' intentions and effects. Carl Mitcham discusses the engineering of objects, implying that physical aspects of man-made objects constitute technology.[10] This author also leans towards the humanities approach to technology and proposes the phenomenological study of the use of artifacts. Christopher Alexander investigates the problems of fitting man-made material forms into daily life contexts.[11] He examines how this fitting is either consciously or unconsciously conducted. Henry Petroski devises the fundamental principle that "existing things fail to do what we expect them to do as well and conveniently and economically as we think they should or wish they would."[12] Thus, the form of objects does not follow their function as it is usually admitted in the classical Product Designers' milieu. The form of objects is usually born out of a perceived failure in existing objects to perform certain needed tasks.

Another line of enquiry concerns the *formal ontological qualities of spatial objects*. For example, a theoretical framework for the robotics of space recognition has stimulated the study of the formal properties of spatial entities, surfaces, wholes and parts, and holes and other superficialities.[13] But, a formal ontology tends to obfuscate the problem, unless a material or substantive ontology is first delineated. According to Stefano Borgo, Nicola Guarino, and Claudio Masolo physical objects are often considered in existing "top-level ontologies" as "belonging to multiple categories,"[14] a situation that generates "confusion, and lack of semantic rigour." This ambiguity is because physical objects can be seen as mass, as extension, as temporal process, or as energy exchange, etc. These general substantive categories of physical objects do not shed light on the differentiating organizational structures that emerge at the level of material artifact. The core properties considered to be invariant *identity criteria* specific to material artifacts, which would differentiate them from the two other geographico-ontological regions, continue to be elusive.

Formal ontology in geography focuses attention on *spatio-material configurations*: that is, the mereo-topological and morphological properties of the region they occupy, and the mereological properties of the elements of which they are composed.[15] Yet, "the formal characterization of identity criteria for entities like artifacts [is] notoriously known as a very difficult problem."[16] "Identity is one of the most fundamental notions in ontology, yet the related issues are very subtle, and isolating the most relevant ones is not an easy task."[17] The philosophers Roberto Casati and Achile C. Varzi examine boundary demarcation in the mereo-topological perspective.[18] Their conclusion is that geographic entities are specified by their boundaries, but these are so vague that there cannot be a precise definition formulated.[19] Henceforth, if material artifacts are also spatial geographic entities, it then follows that their exact nature and boundaries too are inherently problematic.

Thus, studies that do deal with material artifacts have subordinated them to the study of human expressivities, as a function of human geography. Formal ontology tends to obscure the problem, for it functions across hierarchies and levels organizing significances in a "scholastic" manner and thus cannot take the place of material (non-formal) ontologies, which can only be delineated on their own grounds.

We now turn to our ontological delineation of material artifacts. Ron Bennett[20] cites Brian Cantwell-Smith who writes that "material objects—and the material world more generally—occupy what we might call a middle ground, halfway between the physical world, and the intentional world. The resulting median nature of materiality has numerous theoretical consequences."[21] Because we are to return to material ontology, phenomenological analysis, especially Husserl's attention to developing regional ontologies on the principle that being takes on different essential characters according to fundamental regions, seems to offer great promise. However, we are disappointed in that Husserl delineates his fundamental regions as Material Nature, Animal Nature, and Spiritual World.[22] A fundamental clue for the delineation of a material ontological region is found in the writings of Georg Simmel.

> We speak of culture whenever life produces certain forms in which it expresses and realizes itself. . . . But although these forms arise out of the life process, because of their unique constellation they do not share the restless rhythm of life. . . . These forms . . . acquire fixed identities, a logic and lawfulness of their own; this new rigidity inevitably places them at a distance from the spiritual dynamic which created them and which at a distance makes them independent.[23]

This passage provides the insight that Husserl's Spiritual World must be further delineated into "Subjective Culture" and Objective Culture." Once human subjective intentions or expressivities are objectivated in the creation of forms, these objectivations take on a being of their own, different and separate from the human intentions out of which they emerge. These objectivations constitute Objective Culture in that they "stand against" subjective intentions once they have been created, e.g., an economic system with its own structures and dynamic forces, a rain barrel that holds only so much water. These sedimentations of meaning are deposited materially as forces or forms with which one must reckon.

This clue provides the most important reason why it is necessary to distinguish the Geography of Material Artifacts. Once objectivated, material artifacts act back on their makers by: following their own laws, producing their own consequences, and appropriating human life as well as physical nature in their images. It is this detachment, independence, and appropriating efficacy in regards to human intentions that establishes their own being, which makes it necessary to treat material artifacts as comprising their own ontological region.

But we must not forget that human objectivations involve transformations and appropriations of cosmic energy in whatever form it has been crystallized and then re-embodied in more suitable material artifacts, that is, wood into paper, grapes into wine, coal into fuel. This means that material transformations are *objective*, as dynamic forces that *stand against* the workings of nature, because they are transformative of them, that is, the building of dams, the draining of marshes, the use of pesticides, cattle ranching. It is this aspect to which environmental thinkers have been sensitive.

We submit the following principle for the delineation of the Ontology of Material Artifacts: Material Artifacts are objectivations of human expressivities that once expressed exhibit their own being. This being "stands against" both humanity and nature (physical, chemical, biological) due to the fact that material artifacts affect and effect transformations in those two other ontological regions. Material Artifacts are not natural entities, but they always include a physical moment, and they are not human expressions, but they always depend on human expressivity for their existence. Yet, Material Artifacts are not meant as a hybrid or a synthesis of the other two regions: they establish their own webs of relations. Moreover, like all other beings, they participate in various relatively stable organizational systems in the cosmic hierarchy of open dynamics—synergetics.

We now must introduce a crucial distinction, which tangentially allows for flexibility in the treatment of entities such that some of the problems with concepts are eliminated. The material (as opposed to formal) ontology that we advocate is meant as eidetic (phenomenological) and not as substantive (empirical). We do not offer our ontological classifications on the basis of real entities and empirical findings (like Aristotle). As eidetic structures, this threefold system of geographical science allows for the treatment of real entities to be constituted according to the selection of the researcher. For example, a homeowner plants a tree. The tree can be constituted as a natural object and thus can be constituted under the rubric of Physical/Natural Geography. But for different purposes it can be thought of as an artifact and constituted under the rubric of the Geography of Material Artifacts. And, it can be studied under human geography concerning the intentional meaning-contexts involved in purchasing, planting, and maintaining the tree, or even in eventually replacing it with a synthetic one. In our eidetic ontology, we avoid substantive claims, which we think helps to avoid arguing over the substantive whatness of something, for it can be constituted in multiple ways, which serves various purposes.

We have argued that the classification of the geographical sciences needs to include the Geography of Material Artifacts. After reviewing literature concerning substantive material ontology, formal ontology, and eidetic material ontology, we have proposed an eidetic principle that delineates the Geography of Material Artifacts. We have argued for a classificatory system of regional ontology that is constitutive of three regions: Human Geography, Geography of Material Artifacts, and Physical Geography.

Our next step is to promote a "metaphysical science" that provides a common language for the purpose of grasping the interrelation of these three regional ontologies. The proposed synergetic science delineates hierarchical systems-organizations, exhibiting their differences and their relational progressions. Synergetics involves the study of the open dynamics systems in the self-organization of structures and processes.

The Idea of a Synergetics Geography and the Synergetics of Buckminster Fuller

Synergetics is a non-reductive, unifying endeavor in scientific inquiry. "The basic goal of synergetics is to find out the general ideas, laws and principles of self-organization processes in various fields of human knowledge—from natural sciences to the humanities."[24] "The concepts of structure and organization of objects, scrutinized as systems, were devised in science with the purpose to develop a *universal language* adjusted for the description of essentially different systems."[25] The contents of synergetics are the spontaneous, self-organization of structures and processes. Systems-levels involve spatial-temporal processes of progressive organization—the emergent character of the system at a certain relatively stable level cannot be predicted on the basis of, and cannot be reduced to, the behavior of its parts.

The first application of synergetics that we advocate is to develop a common language that can exhibit the interrelation of the three fundamental ontological regions of geography, which are three differently organized types of systems based on their respective ontological principles. These systems, however, entail a hierarchy of progressively emergent organizations at relatively stable levels of physis, bios, artifact materiality, society, and culture. Physical/Natural Geography primarily concerns the relatively stable systems of physis and bios with their various self-organizing components. The Geography of Material Artifacts primarily involves moments of physis, bios, and human values conveyed through artifactual forces, such as that inscribed in coins and money exchange procedures, in baseball games and bats, in library systems and books, or in the manufacture of dog food. Human Geography concerns human populations as societies and cultures through the expressivity of human meanings. This region entails moments of physis, bios, artifactual materiality, society and culture. These systems-levels and the ontological delineations interrelate in complex ways and there needs to be a way to treat those interrelations geographically.

Because of the emphasis we have placed on the ontological region of material artifacts, we focus specifically on the synergetics of the world-renowned architect-mathematician and philosopher, Richard Buckminster Fuller.

> When I say *artifact*, I mean any participation using the principles of nature to reassociate these principles for a specific purpose.
>
> The methodology of my artifact revolution is quite simple. Taking nature's cue, I determined that I must commit myself to solving problems by artifacts: what I call reforming the environment. . . . The function of what I call design science is to solve problems by introducing into the environment new artifacts, the availability of which will induce their spontaneous employment by humans and thus, coincidentally cause humans to abandon their previous problem-solving behaviors and devices.[26]

It is through his interest in technology that Fuller enters into synergetics. He devised his theoretical premises in *Synergetics* and on its basis built practical constructions, the well-known geodesic domes.[27] But *technology* is conceived as the fundamental way of the Universe. Human intentions are a moment of this ubiquitous technology in that as an aspect of Universe they participate in a circular causality of "makings"—humans are unique technologies of Universe that can introduce technologies into being in a knowing way. Thus, Fuller sets out to discover a common framework for technology—"eternally regenerative integrity of Universe."[28] Thus, human, material artifact, and physical/natural world are all linked by this common conception.

Fuller ascertains that, cosmologically, there is no such thing as solid matter, which includes the earth's crust and all that it supports, which traditionally constitute the protocol of substantive geographical studies. The entire universe is nothing but energy waves, entirely and solely characterized by different magnitudes and frequencies. Those waves crisscross and at moments concentrate associatively. The resulting sensible *physical* build up thus gives the impression of solidification into "matter." Those same waves may disassociate from each other and run off in radiation quanta, towards other new associative encounters, deterioration of solidification. It is in this constant double and non-simultaneous movement of synergetic integration and diffusion of energy waves that apparent solid, liquid, or gaseous structures emerge as temporary, never permanently stable patterns.

Fuller sheds the concept of space, except in terms of relatively stable systems. "There is no universal space or static space in Universe. The word *space* is conceptually meaningless except in reference to intervals between high-frequency events momentarily constellar in specific local systems."[29] "Space is the absence of events, metaphysically. Space is the absence of energy events, physically."[30] "We have relationships—but not space."[31] As a relational theory of spacetime, Fuller follows in the tradition of Leibniz and eschews the Newtonian substantival theory of spacetime.

It follows then that "there are no solids or particles—no things."[32] "Physics has found no 'solids.'"[33] "The synergetic behavior of structures satisfactorily explains as discontinuous that which we have in the past superficially misidentified as 'solid.'"[34] "We have now learned . . . that there are no . . . continuous surfaces, only Milky Way-like aggregations of remotely interdistanced atomic events. Nonetheless, society keeps right on seeing, dealing, and superficially cerebrating in respect to 'things' called 'solids' or 'matter.'"[35]

"Since the word *space* as a noun, misleadingly implies properties that are altogether lacking,"[36] it seems that ontological geographies deal with mere illusions. But then we would be misapplying concepts across system-levels, if we were to eliminate categories such as space from ontological geographies. What we must recognize is that for our purposes of envisioning Synergetic Geography, "space" is not a category with which we will employ. Our fundamental categories will be energy and synergy. "Energy relates to differentiating out subfunctions of nature, studying objects isolated out of the whole complex of Universe."[37] Thus it is the manifesting outcomes of energy with which the ontological geographies deal. "Synergy represents the integrated behaviors instead of all the differentiated behaviors."[38] This means that Synergetic Geography will not deal with the ontological outcomes of energy, but with energy itself as it manifests in systems of synergy.

"A system is the first subdivision of Universe into a conceivable entity separating all that is nonsimultaneously and geometrically outside the system."[39] "Systems always divide all Universe into three principal parts: the system itself; all Universe outside the system (the macrocosm); and all Universe inside the system (the microcosm)."[40]

> The . . . three-way division can be expanded into five zones. All Universe outside the system considered is divided into (1) the clearly irrelevant macrocosm zone and (2) the twilight macrocosm zone of tantalizing possible relevance. The next zone is (3) the system itself; clearly relevant and tuned-in, it convergently-divergently divides all Universe into its macro-outsidedness and its micro-insidedness irrelevancies. The microcosmic insidedness is divided into (4) the twilight microcosm zone of tantalizing possible relevance and (5) the clearly irrelevant microcosm zone.[41]

"Universe is a technology—the most comprehensively complex technology."[42] The local technologies, humans and their artifacts, are among all of those specifiable localized scenarios, each at its specific systems-level organization and with its corresponding inwardly and outwardly energy exchanges.

Humans are at the most progressive relatively stable systems-level in evolutionary dynamics whose main differentiating feature is the ability to intentionally *produce* (etymologically meaning to "*draw forward*") similar as well as different physical and metaphysical phenomena or events. In Fuller's exact wording:

> Of all the subcosmic, integrally interpatterning complexes that we know of in our Universe, there is no organic complex that in any way compares with that of the human being.[43]

> Humans are the most complex local technologies.[44]

Humans are contextualized as another of the multiple *local technologies* within Nature's Synergetics, but their complexity of systems-organization endows them with a unique function in regard to other local technologies. "Wherever they occur in Universe, human beings may be introduced as means of coping metaphysically with the most complex kinds of local Universe problems, so that each one of us is where the problem-solving Universe is being transacted."[45]

So, in Synergetics, the "it" of previous reference is now treated, whether human, physical/natural, or material artifact, as a complex field of cyclic waves with inward and outward frequencies. The "it" is the individual experience of every cosmic entity, periodically determined in its frequency and in its magnitude. This characterizes both the physical and the metaphysical aspects of our observing faculties and the observed phenomena. The "it" may or may not manifest to humans perceptually or conceptually, because of the respective relatively limited, specifically bounded physical and metaphysical scanning capacity of earth's human inhabitants.

This common language of synergetics helps us to develop the Synergetic Geography that forms an ecologization of our threefold ontological geographies at deeper systems-levels. Whether we are dealing with a system that manifests as physical, human, or material artifact in our ontological concerns, in Synergetic Geography, they are constituted as local technologies—a perduring complex of relatively stable energy exchanges. The systems of the Universe result in local technologies, that is, "makings" that manifest to us ontologically. Fuller's "deep ecology" is that problems occurring at the ontological level cannot be adequately treated by remaining at that level. Those problems can only be adequately dealt with by a synergetic ecology, the interrelationships of local technologies, which does not deal with ontological entities, but rather synergetic systems and their energy flows.

The fundamental dynamic principle of the Universe is the simultaneous inwardly and outwardly motions. "Energy transforms and trans-shapes in an evolutionary way,"[46] converging and dispersing in a covariably coupled frequency and wave.[47] "Two different energy events . . . cannot pass through the same point at the same time. When one energy event is passing through a given point and another impinges upon it, there is an *interference*."[48] "Everything we call structure is synergetic and exists only as a consequence of interactions between divergent compressional forces and convergent tension forces."[49] "A structure is a system of dynamically stabilized self-interfering and thus self-localized and recentering, inherently regenerative constellar association."[50] "*Regenerative* means local energy-pattern conservation. *Constellar* means . . . locally co-occurring events dynamically maintaining their interpositioning."[51] Microconstellations manifest as forms of matter—water, wood, cheese, aluminum foil.

All the Universe components are involved in an endless exchange of energy. Within this Universe, a human being's life is a *double interference mode* in the exchange process: both metaphysical and physical energy investment and divestment. Through human intervention, the Universe is thus increasingly filled with and made of various types of energy waves of different magnitudes and frequencies. These trajectories, at certain times, bundle up in perceptible forms of clearly bounded "spatial entities," material artifacts that are bound to an everlasting complementary growth and decay.[52]

The dynamic actions of energy motions manifest as well as multi-interference *between* local technologies. The eidos that we proposed for material artifacts as *standing against* human intentions and *standing against* the natural/physical domain has its root in the synergetics of interference. Interference does not mean obstruction or destruction, because it just as well applies to facility, construction, and self-stability. The *quality* of the relations on the ontological level between human, physical/natural, and material artifacts, needs to be linked to synergetic dynamics of the interfering energy events. Synergetic systems exhibit shapes, limits, and boundaries, due to the dynamics of energy flow.

> The regenerative patterns of structural events may be described as *constellar* because their component events interfere tensively—and gravitationally—in high-frequency, dynamic, self-regenerative patternings which only superficially seem to stand together as 'static' structures.[53]

Systems retain their boundaries through tensegrity, which is the principle of structural-relationship providing "the ability to yield increasingly without ultimately breaking or coming asunder."[54] This is what could be labeled as the universal non-violatable condition of outward and inward flows. Radiation distributes energy systems outwardly. Gravity counteracts radiation; it is progressively and centrally focussing.

> An aggregate of "falling ins" is a body. . . . The . . . transition from being an aggregate entity to being a plurality of separate entities is precession, which is a "peeling off" into orbit rather than falling back in to the original entity aggregate.[55]

Every local technology is thus always energetically bound both inwardly and outwardly. At the ontological level we can link the dead fish in a lake with the lack of oxygen supply in the water. This lack of oxygen occurs through the growth of certain microbiotics caused by certain pollutants entering soil run-off through the waste storage by-products of a factory. We can link these events ontologically through logic, trial and error, and empirical observation. However, synergetics allows for a deeper understanding through linking the energy systems of local technologies, including human intentions, material artifacts, the natural chemical processes, and the biotics of aquatic life all in one language. Thus, synergetics allows for the possibility of reconfigurating the energy flows by making adjustments throughout the system, which may result in various "technological" adjustments that can affect all three ontological domains.

When *intentional interference* in local scenarios is a more or less systematized process peculiar to the human mind, then the process is of *Design*. Design of material artifacts should be understood as the human outward interpretation and implementation of the synergetic laws of the Universe into physical and metaphysical forms. Fuller's theory could be termed, the "eternal cosmic design"—the family of generalized principles or interrelationships present in the Universe for designing systems. In the common human expression, the word *design* "is used in contradistinction to random happenstance. Design is intellectually deliberate. Design means that all the components of the composition are interconsiderately arranged" in new structural transformations.[56] Design is a conceptual process, also never static, like the Universe within which the process is triggered and evolves. It is a never-ending thinking process peculiar to the human project. "Tools, artifacts, and all humanly contrived extracorporeal facilities designed by mind are always special-case, brain-processed, physically sensible phenomena."[57] Conceptualizations are formalized and crystallized into material structures, as are other local technologies, transformed and fixed to specific purposes with special corresponding learned skills and techniques. This process and its results constitute the "material culture" of a given human collectivity.

All the obtained material structures are intrinsically spatial fixes of criss-crossing trajectories of energy. Therefore, we argue that the whole process of artifact design and production, the entire material culture of a given human collectivity is nothing else but a conceptual cosmic energy appropriation, apportionment and containment. This is done through appropriately devised skills and techniques, and the universal energy is tapped into structures, shaped into material artifacts, which are then allotted to certain human sets of life purposes, whether individual or collective. We also propose that wherever a human is, he is constantly conceptualizing and forming mental and physical energy-bonded structures with which he interacts and with which he, forcibly or through peaceful negotiation, brings other human and earth entities to interact. Eventually, material artifacts of different kinds and of different sizes are built and gradually spread over the landscapes, in closer tight proximity of other "energy fixes."

Let us illustrate the sense of relevant processes associated with a fix. An earth slab occupied and delineated as one's own; a leaf carefully grown on this particular slab and harvested with a knife made out of the same slab's minerals; chopping, drying, and grinding equipment for the leaf's transformation into a special medicine to be made in the form of tablets, powder, liquid; laboratory buildings and instruments for active extraction of ingredients out of the leaves; forming and conditioning of obtained extractions; distribution, reception, and intake of the prescribed medicine; the overall social and material organization for full and relatively large production of leaves, medicine, and corresponding treatment. These processes are all different physico- and sociocultural features designed out of a particular earth slab "territory." They are all artifactual territorial expansions of the one primary core energy fix, the composite of a particular medical ingredient, which undergoes a succession of designings to serve specific human purposes.

Structural transformations are conceived by the human mind through modeling. Fuller defines modelability as "topologically conceptual in generalized principles, independent of size and time; ergo, conceptual modelability is metaphysical."[58] The "conceptual formulation is inherently empirical and as such is always special case sizing and always discloses all the physical characteristics of existence in time."[59] The outcome of this double process is that "model is [conceptual] generalization; form is special case" realization.[60]

> Human awareness is conceptually initiated by special-case otherness observability. Humans conceptualize, i.e., image-ize or image-in, i.e., bring-in, i.e., capture conceptually, i.e., in-dividualize, i.e., systemize by differentiating local integrities from *out* of the total, nonunitarily conceptualizable integrity of generalized Universe.[61]

All humans, in all cultures, "image-ize" in one way or another their perceptible local technologies, whether these be the humanly body as their immediately experienced territory or the externally perceived territories, including man-made material artifacts and their assigned location. The image can be spatially represented. In Western culture, the eighteenth-century Swiss mathematician, Leonhard Euler, and before him the Greek philosopher and geometrician, Euclid, endeavored to prescribe rules of space description, rules of *topology* that have now become norms in any spatial detailing. Image-izing is precisely what geographers, engineers, architects, and other designers do according to their respective concerns. The fundamental image representation of geographers is maps. The classic image representation of Western geographers is two-dimensional maps. They are generally drawn with points, lines, and planes.

> Such self-deceiving misinterpretations of experiences have been introduced by education into human sensing and traditional reasoning only because of humanity's microstature and microlongevity in respect to the terrestrial environment and geological time. . . . Humanity has been misinformed or underinformed by the spoken-word-relayed inventory of only popularly explained, naked-eye impressions of local environment experiences as they have occurred throughout millions of years prior to humanity's discovery and development of instrumentally accommodated, macroscoped and microscoped exploration of our comprehensive environment.[62]

Augmenting Euler's triad of vertexes (points), faces (planes), and edges (lines), Fuller proposes a Synergetics topology essentially based on *Ins*, *Outs*, and *Interrelationships* among all Universe systems.[63] Instead of the only three assumed fundamental structural elements, there are thirty-two minimum aspects in the Universe.

Synergetic Geography involves an image-izing topology of the energy events constituting synergetic systems. Unlike image-izing for design—architecture, engineering, etc., Synergetic Geography's first task is to map and model the relevant synergetic systems concerning sectors of the earth's landscapes. Synergetics topology would be associated and correlated with the mappings and modelings of the ontological geographies. Its purpose is to complement them with models and maps that convert the ontologically differentiated relevancies into synergetic systems of energy flows, which allows for a deeper description of ecological relations. From these complementary functions of mapping and modeling, Geography contributes its special expertise in the quest to "make" a better world.

The human species is increasingly intervening globally in the everlasting Universe "tensegrity game," (that is, each of the Universe components and elements gravitationally attracting others which, in compensation, are in permanent reactive tension). But it is precisely the "global universal game" that, as so rightly indicated by R. Buckminster Fuller, is not well understood and consequently badly played by us all to our own detriment.

The Scope of the Geography of Material Artifacts

The Geography of Material Artifacts can concern man-made things the size of a stapler, a sheet of paper, a room, a house, a lawn, a bounded community, a territorial state, or a region—to give some sense of its possible applications. Mappings of garbage dumps, hospitals, railroad tracks, swimming pools, chain-linked fences, gasoline stations, aerosol cans, the contents of a suburban garage, just about any imaginable types of material artifacts in any types of combination could be subject to geographical mappings. Certain densities, volumes, and weight of material artifacts are important to certain types of research. It is not that studies do not already do this, but what is of great importance is to treat these material artifacts qua material artifacts; that is, not to reduce them to either

Physical Geography or Human Geography; even worse is not to acknowledge their own ontological being. Certain kinds of material artifacts, economic systems, bureaucratic practices have long been studied as energy moving forces. However, it is now important to treat all artifacts, including material artifacts, even seemingly the most trivial, from the standpoint of their own being such that they *stand against* and constantly *interfere* with intentions—facilitating, thwarting, transforming, and molding them—and *stand against* and *interfere* as well with the cosmic forces by transforming ecological conditions. The mere geographical mappings and modelings of material artifacts may be useful in suggesting further studies that provide explanation or motivate a critical valuation for change or maintenance of the status quo.

The Geography of Material Artifacts shows it worth even more in relation to the other two ontological geographies. Mappings could exhibit human intentions in relation to those material artifacts constituted as useful in fulfilling those intentions. The qualities in the relations between subjective intentions and the ontological character of material artifacts must be ascertained. The Geography of Material Artifacts ascertains the spatial relations of the artifacts, which in these studies would be related to the relevant human intentional contexts. In relation to Physical/Natural Geography relations between artifacts and cosmic elements/natural objects are especially important for ecological maintenance and improvements. Mapping Floridian highways and roads to determine their relationship with the wetlands that they circumscribe and cross. Mapping the rising number of community golf courses in relation to the water tables of a region. Mapping the number of birdfeeders in a neighborhood.

Obviously, a triadic ecological mapping or modeling of the three ontological geographies can be most beneficial for description of successes and failures and prescriptions and proscriptions concerning technological transformations. The relations between human intentions, human habitats, and wilderness might be subject to a geographical study. Again, it is the recognition of the ontological delineations such that the material artifacts are promoted on their own terms that properly frames the problem. Once the role of the material artifacts is ascertained in such a study, then prescriptions and proscriptions can better be articulated. These evaluations will not be as fruitful of guides if the being of the material objects is not properly distinguished from human expressivity. For example, problematic situations could be on the side of the artifact itself, or merely the human intentions, or both. Geography can map each alone; and even better, both together.

The relation between the Geography of Material Artifacts and Synergetic Geography is fundamentally crucial. Humanity lives through its artifacts and human material interventions and objectivations potentially affect every other entity as well as entity-relations. But affects and effects are not always recognized or indicated at the level of perceivable reality. Certain indirect perception e.g. the reading of instruments, and indirect conceptual methods are necessary to ascertain situations. Synergetics provides a language that allows for the relevant systems to be related from a deeper standpoint. Synergetic Geography must provide topologies, maps and spatial models, of the systems, their interferences, and their tensegrity in the radiational and gravitational exchange of energy. The topology of organizing systems provided by Synergetic Geography must be correlated to the mappings of the ontological geographies. Design involves conscious transformations initiated by humans, which means material artifacts are a moment of the living process. Thus, it is most important that material artifacts be properly mapped and delineated by the Geography of Material Artifacts prior to and after the initiated transformation. This geography must correlate with the mappings of Synergetic Geography before and after in order to evaluate the process at the deeper levels of energy exchange.

The differentiated ontologies respect the hierarchical organizations of systems through their delineation. Physis, bios, human society, and cultural organizations all have organizational differences. It was not the goal of this paper to distinguish these levels of evolutional progression. But our ontological geographies assume their differences on their limited basis of differentiation. Synergetic Geography also respects these differences, but not ontologically; it respects these differences through the disclosure of the energy systems-organizations that they manifest. The ethical axiological ramifications for design decisions are best served through the complementary relation of both forms of geography. So, after its own inclusive descriptive mappings and modelings, the Geography of Material Artifacts must enter into a dialogical relation with the other geographies in order to fulfill a collective purpose and goal.

Concluding Remarks

The purpose of this chapter has been threefold. Our first goal has been to provide a theoretical framework for the delineation of the Geography of Material Artifacts. This geography has been contextualized in a threefold delineation of what we have labeled as ontological geographies. We then provided a discussion of synergetics in order to gain a common language that links all levels of being in one common language—the language of systems. We specifically discussed the synergetics of Richard Buckminster Fuller, because his Synergetics most germanely captures a language of "making," of technologies, that links together all processes and structurings of the Universe through the flows of energy and the resultant manifestations. We have conceived a Synergetic Geography that provides topological mappings and modelings of energy flow in synergetic systems. Finally, we outlined the scope of the Geography of Material Artifacts and explicated the complementary relationship between the ontological geographies and Synergetic Geography.

It has been our goal to offer a framework that enables a more realistic account of the structures containing and conveying the entire cosmic energy. More than simply accounting for perceivable entities, especially the man-made energy enclosures, material artifacts, through a synergetic understanding of design, we will be better equipped to deal with the growing problems of our own "makings." The problem of designing would then be greatly aided by geographical prescriptions and proscriptions. Cognitively tapping cosmic energy, and then technically fixing "it" into appropriate and useful mental, intellectual and material artifacts ("what"). There should follow a socially responsible allotment of those energy contained artifacts ("where") and a wiser individual and ethically regulated appropriation of such energy bound artifacts for daily life survival ("why").

Notes

Acknowledgments. François-X. N. I. Nsenga: My heartfelt thanks to Gary Backhaus and to my daughter, Ak'ingabe M. Guyon. Gary Backhaus: Many thanks to François-X. N. I. Nsenga for his brilliant ideas and his cordial collaboration.

1. Jacques Grand'Maison: *Pourquoi sombrons-nous si souvent dans la démesure?* "Les grandes conférences Desjardins 10, Programme d'études sur le Québec," McGill University, le 13 mars 2002.

2. Free interpretation of: *Le village; Le terroir irrigué; Le finage d'alentour.* See Jacques Berque, "Vie sociale et variations de mode et de densité," in *Sociologie des mutations*, edited by Georges Balandier (Paris: Éditions Anthropos, 1970), 130-31.

3. Basarab Nicolescu, *Manifesto of Transdisciplinarity*, trans. Karen-Claire Voss (Albany: SUNY Press, 2002).

4. Richard Hartshorne, *The Nature of Geography* (Lancaster, Penn.: Association of American Geographers, 1939), from *Chapter 12: Conclusion: The Nature of Geography.* Text converted by Edwin Chow and Russel Taylor. Created on 1999.11.21. Last revised 2000.4.1. by Kenneth E. Foote. http://www.utexas.edu /depts/grg/ustudent/gcraft/fall99/chow/Hartshorne/chapter12.html, accessed October 2002.

5. Bill Manger, "Diversity and Globalization (8/23). Overview of Geography," in *Outline of Course Notes* (Tempe: Arizona State University, Fall 2000).

6. Among the few rare references: Thorstein Veblen, "Conspicuous Waste," in *The Theory of the Leisure Class.* (New York: Mentor, 1953, 1899). Jean Baudrillard, in *Pour une critique de l'économie politique du signe* (Paris: Gallimard, 1972).

7. Tim Dant, in *Material Culture in the Social World* (London: Open University Press, 1999).

8. Risto Hilpinen, "Artifacts," in *Stanford Encyclopedia of Philosophy*, edited by Edward N. Zalta (Stanford, Calif: A Publication of the Metaphysics Research Lab, 1999).

9. Randall R. Dippert, Artifacts, Artworks, and Agency (Philadelphia: Temple University Press, 1993), 23-37.

10. Carl Mitcham, *Thinking through Technology: The Path between Engineering and Philosophy* (Chicago: The University of Chicago Press, 1994).

11. Christopher Alexander, *Notes on the Synthesis of Form* (Cambridge, Mass.: Harvard University Press, 1964).

12. Henry Petroski, *The Evolution of Useful Things* (New York: Vantage Books, 1994), 22.

13. Roberto Casati and Achile C. Varzi, *Holes and Other Superficialities* (Boston: MIT Press, 1994).

14. Stafano Borgo, Nicola Guarino, and Claudio Masolo: "Stratified Ontologies: The Case of Physical Objects," *European Conf. On AI (ECAI96)*, Budapest, August 1996. Another version of the same is entitled: *Towards an Ontological Theory of Physical Objects.* (Italy, National Research Council, LADSEB-CNR, 1996).

15. Mereology is the study, in formal ontology, of the relationships between wholes, their parts, and the boundaries that constitutes parts as discrete entities. Mereotopology is the study of emplacements or positions of wholes, parts, and boundaries.

16. David Wiggins, *Sameness and Substance.* (Oxford: Blackwell, 1980). Peter Simons, *Parts: a Study in Ontology* (Oxford: Clarendon Press, 1987). Borgo, Guarino, and Masolo: *Towards An Ontological Theory.*

17. Nicola Guarino and Christopher Welty, "Identity, Unity, and Individuality: towards a Formal Toolkit for Ontological Analysis," in Werner Horn (ed.), *ECAI-2000: Proceedings of the 14th European Conference on Artificial Intelligence* (Berlin: IOS Press), 219-23.

18. Roberto Casati and Achile C. Varzi, "Spatial Entities," in Oliviero Stock (ed.) *Spatial and Temporal Reasoning* (Dordrecht: Kluwer, 1997), 73-96.

19. Achile C. Varzi, "Words and Objects," forthcoming in Andrea Bottani, Massimiliano Carrara, and Daniele Giaretta (eds.), *Individuals, Essence, and Identity: Themes of Analytic Meaphysics.* (Dordrecht: Kluwer Academic Publishers). See Achile C. Varzi, "Vagueness in Geography," in *Topoi* 20, (Dordrecht: Kluwer Academic Publishers, 2001), 119-30. See David M. Mark, Barry Smith, and Barbara Tversky, "Ontology and Geographic Objects: An Empirical Study of Cognitive Categorization," (Heidelberg: Springer Verlag, 1999), from http://link.springer.de/link/service/series/0558/bibs/1661/16610283.htm, accessed October 2002. See Roberto Casati and Achile C. Varzi, "Ontological Tools for Space and Geographic Representation," in *Workshop on "Formal Models of Common-Sense Geographic Worlds"* (San Marcos, Tex: October 30, 1996). See Amie L. Thomasson, "Geographic Objects and the Science of Geography," in *Topoi* 20 (Dordrecht: Kluwer Academic Publishers, 2001), 149-159. See Brandon Bennett: "What is a Forest? On the Vagueness of Certain Geographic Concepts," in *Topoi* 20 (Dordrecht: Kluwer Academic Publishers, 2001), 189-201. See Barry Smith, "Fiat Objects," in *Topoi* 20 (Dordrecht: Kluwer Academic Publishers, 2001), 131-148.

20. Ron Bennett: *Critical Approaches to Culture.* In *Communications + Hypermedia*, a set of resources on the World Wide Web. http://www.eciad.bc.ac~rburnett/weekseven.html, accessed October 2002.

21. Brian Cantwell Smith, *On the Origin of Objects* (Boston, Mass.: The MIT Press, 1996).

22. Edmund Husserl, *Ideas Pertaining to a Pure Phenomenology and to a Phenomenological Philosophy: Second Book*, trans. R. Rojcewicz and A. Schuwer (Dordrecht : Kluwer Academic Publishers, 1989).

23. Georg Simmel, *On Individuality and Social Formal : Selected Writings*, ed. and with an introd. Donald N. Levine (Chicago : The University of Chicago Press), 375.

24. Michael Bushev, *Synergetics: Chaos, Order, Self-Organization* (Singapore: World Scientific, 1994), vii.

25. Bushev, *Synergetics*, 20.

26. R. Buckminster Fuller, *Cosmography: A Posthumous Scenario for the Future of Humanity*, Adjuvant, Kiyoshi Kuonmiya (New York: Macmillan Publishing Co., 1992), 8.

27. R. Buckminster Fuller in collaboration with E. J. Applewhite, *Synergetics: Explorations in the Geometry of Thinking* (New York: Macmillan Publishing Co., 1975) and

Synergetics 2 Further Explorations in the Geometry of Thinking (New York: Macmillan Publishing Co., 1979).

28. Fuller, *Synergetics* 107.00, 4.
29. Fuller, *Synergetics* 526.01, 272.
30. Fuller, *Synergetics* 526.03, 272.
31. Fuller, *Synergetics* 526.04, 272.
32. Fuller, *Synergetics* 240.08, 63.
33. Fuller, *Synergetics* 524.03, 269-70.
34. Fuller, *Synergetics* 525.01, 271.
35. Fuller, *Synergetics* 525.02, 271.
36. Fuller, *Synergetics* 524.02, 269.
37. Fuller, *Synergetics* 105.00, 3.
38. Fuller, *Synergetics* 105.00, 3.
39. Fuller, *Synergetics* 400.02, 95.
40. Fuller, *Cosmography*, 124.
41. Fuller, *Cosmography*, 124.
42. Fuller, *Synergetics* 311.02, 85.
43. Fuller, *Synergetics* 311.01, 85.
44. Fuller, *Synergetics* 311.02, 85.
45. Fuller, *Synergetics 2* 311.14, 63-64.
46. Fuller, *Synergetics 2* 223.89, 51.
47. Fuller, *Synergetics 2* 223.90, 52.
48. Fuller, *Synergetics* 517.01, 252.
49. Fuller, *Synergetics 2* 792.30, 172.
50. Fuller, *Synergetics* 600.03, 314.
51. Fuller, *Synergetics* 600.04, 314.
52. Fuller, *Synergetics 2* 1053.80, 432.
53. Fuller, *Synergetics* 510.08, 241.
54. Fuller, *Synergetics* 700.011, 372.
55. Fuller, *Synergetics* 131.00, 11.
56. Fuller, *Synergetics 2* 792.01, 171.
57. Fuller, *Synergetics* 326.40, 76.
58. Fuller, *Synergetics 2* 900.11, 198.
59. Fuller, *Synergetics 2* 900.12, 198.
60. Fuller, *Synergetics 2* 900.31, 198.
61. Fuller, *Synergetics* 1006.14, 621.
62. Fuller, *Synergetics 2* 1001.26, 362-63.
63. Fuller, *Synergetics 2* 269.00-269.07, 51-2.

Chapter 7

A Contextualized Science and the Changing Landscapes of India: A Case Study of Science as a Graft

Deepanwita Dasgupta

Lead us from Unreal to Real, from Darkness to Light, from Death to Immortality.[1]

Prologue: The Puzzle of Joseph Needham[2]

My concern, in the following chapter, is to work out the problems of a non-contextualized science and the consequent dilemma of development, taking the history of Indian science as a special case study. To situate this problem with reference to the Indian history, I shall begin with a particular formulation of the Needhamian question that I will call the 1660 question. Needham's question referred to the general conditions that prevailed in the seventeenth-century Europe and raised the problem why was it that the scientific revolution took place in Europe and not in China. This is indeed a most curious puzzle, which has called forth a number of answers, none of which, to my knowledge, has been universally accepted. I begin here with a specific version of that question and simply ask why around 1660, the Indian rulers did not take a more active interest in promoting the bodies of knowledge that are today known as "scientific" when they apparently had an abundance of resources to do so. My tentative reply to this question is that this was due to an extreme form of self-satisfaction, which is reflected in their documents of state.

The 1660 question is modeled, of course, on the 1492 question that Giere raises in his interchange with Gopnik over the similarly perplexing issue of the (cognitive) origins of scientific knowledge.[3] It brings into focus the puzzling way scientific knowledge makes its late entry into the Indian subcontinent under the British rule, having earlier been neglected by its native rulers, who failed to discern its future importance. 1660, we remember, was the year of the foundation of the Royal Society in London. Correspondingly in Indian history, it was the reign of Shah Jehan, the most prosperous ruler of the Mughal Empire who ordered the Taj Mahal to be built. There was no great disparity in wealth or in quality of life at that point of time between the two continents. If anything, the disparity existed in the reverse direction. The enlightened policy of the great Mughals was to cherish and to patronize all available knowledge forms within their kingdom, which we see in their liberal support for literature, painting, and architecture. Yet for all its prosperity, this period of Indian history (that keeps several Art History Departments busy today) showed little awareness or appreciation of what we understand as scientific knowledge. A tentative explanation of this remarkable phenomenon could be that for whatever reasons such forms of knowledge were not foreseen as necessary. Yet physically speaking, the new knowledge forms often existed side by side with the old. Since the late sixteenth century, a trickle of scientists, engineers, and medical men had started coming to the Indian sub-continent from Europe. With time, the trickle turned into a steady flow leading to an eventual repackaging (or partial loss) of the entire indigenous knowledge system. To cite a few anecdotes of such early encounters, Thomas Roe, the ambassador of King James I, visited Jehangir's court in 1615 and obtained, apparently without difficulty, a trading agreement for the young East India Company. Clearly, the new knowledge forms were allowed to come in and were even used in times of emergency, but they were not perceived as sufficiently striking or important to warrant their introduction into the existing indigenous bodies of skills or knowledge. As a result of this 1660 negligence, the true interaction with European sciences started late in the eighteenth century, in the colonial period, when after Wood's Dispatch (1854), the first Indian universities modeled in the British style were established in Calcutta, Bombay, and Madras. By then, of course, the power equations had entirely changed. India was no longer a politically independent country, its wealth was gradually draining away to enrich the British treasury, and most of its famous jewels had already become the property of the British Crown. It had lost the chance of engaging with modernity and its main tool, science, as an independent society according to its own needs and concerns.[4] It will always remain a matter of speculation what would have been the trajectory of the history of science in India, had it been prompted by its native rulers and not by its colonizers. The fact, however, remains that when at the end of the nineteenth century, India's engagement with European sciences started within the demands set by the Raj,[5] it started off at a disadvantage where India was powerless to accept or reject any policy according to its own needs. The 1660 negligence had indeed shaped history.

The Changed Landscape: Janus Face of Development

What India faces today is the problem of how to construct a meaningful account of its 140-year-old engagement with science and to work out a beneficial face of "development" through this knowledge. The problematic relationship between science and development is beginning to be felt all over the world in different forms, but India can be accepted as one of its extreme case studies, given the colonial beginnings of its science, its violent introduction to modernity, and its immediate post-independence need for development. Science and its close cousin, technology, hold multiple promises for India today by raising in turn both high hopes and serious anxieties. On the promise side, it hopes to bring into existence a better infrastructure, an enlightened way of life, and, in general, a wonderful key that will solve most of its problems that it has accumulated as a result of its colonial past. On the flip side, it also gives rise to the grave anxiety that once such technologies are accepted they will sweep aside the society's age-old cultural fabric and patterns of human relationships, leaving poorer substitutes in place. The results of several such engagements have already brought home a sobering message. More often than not, "developments" have brought into existence an unacceptable way of life, leading to much pollution of the environment, changing the landscape forever. The newly developed suburbs that are springing up around every Indian cityscape, filled with expensive high-rise apartments, are cases in point. They have, in the recent decades, practically induced a culture of reinforced concrete as the primary city building material, leading to a gradual disappearance of the older, traditional, and locally available materials. The older materials are on their way out because they require more frequent maintenance and are less durable, despite the fact that they consume much less electricity to keep the building interiors cool and are aesthetically more appealing. Such changes are being labeled as progress and are accepted "with a shrug" as the inevitable price of development. If we turn our attention from urban development to the power sector, the same cautionary message is repeated even louder. For more than two decades, there has been a raging controversy over the fate of the Sardar Sarovar Dam on the Narmada River, a project that promises to generate large amounts of electricity for the state of Gujrat. It also promises to bring in large number of jobs to the state. However, if it is implemented—which, after the Indian Supreme Court lifted its four-year ban in October 2000, it seems that it will be—it will result in the flooding of a huge part of the river valley that spans the adjacent states of Madhya Pradesh and Maharashtra. The flooding will uproot a substantial tribal population for whom this land has been home since time immemorial. It is now estimated that the displaced families will number well over 100,000, and the state's plans to rehabilitate or resettle them clearly fails to address the magnitude of the need. The *Narmada Bachao* (Save Narmada) movement launched by the well-known social activist Medha Patkar over the fate of these tribal "oustees" has grown into a movement of international proportions. The making of a dam seems to have become a battlefield of issues involving human rights, degradation of the environment, and what is really meant or understood in the use of the word "development."

The damming of rivers had earlier been perceived as the solution to a number of problems of a developing economy, like agriculture, clean water and electricity, but there is an increasing consensus now that the costs often outweigh such projected benefits. Irrigation schemes or hydroelectric generation projects often fail to satisfy expectations when reservoirs become useless as they fill up with sediments, giving rise to large amount of greenhouse gases and causing large scale disruption of aquatic ecosystems. Indeed, looking at the project performance of dams all over the world, there is reason to doubt whether such investments can really be called investments in clean electricity. Cases like this push us toward the larger and more troublesome question of an acceptable face of modernity in India, and, also, to the more troubling, deeper issue: does India need to engage with this modernity completely empty-handed, while its classical tradition and age-old ways of life are rendered irrelevant by this new knowledge? Does it have to begin anew from everywhere to switch over to a scientific worldview, as some scholars have recently felt?[6] How shall development be understood, and what is the role that will be played in this development by India's science professionals? The professional scientific community in India has existed ever since the first decades of the twentieth century, but frequently its overall stance on India's future self-image seems not to be too clear. Indeed, the answers regarding the aim of science and what science ought to achieve in Indian contexts seem to be largely borrowed from societies elsewhere that face altogether different concerns. A body of theoretical thinking about sciences, which relates those forms of knowledge to the local settings and existing classical traditions, seems to be clearly lacking at this point. The task at hand is to decide what kind of modernity satisfies India's contextualized needs. Is it possible to find a passage between India's classical ideals and the new task that it is facing today? Does the older tradition of the country include ideals of happiness or freedom from want, and, if it does, can those ideals be made to converge (or at least overlap) with the modern ideals of life, liberty, and the pursuit of happiness? I think the answer to this question can be a qualified yes, because, as we shall see, those concepts admit of reinterpretation. On the other hand, it is becoming increasingly clear today that the notion of a beneficial science needs concepts like an eco-system or a "network" where man and her/his environment is seen as an organic unity, each influenced by the fate of the other. Science does not lose its leading role in such an engagement but allows itself to be placed into contexts by being informed by other systems of knowledge. A contextless science spells danger to any society, but is especially dangerous to a developing society that is in transition, because it allows science to function unexamined.

This brings us to the issue as to what kind of modernity would be acceptable to India's needs. While I do not have a ready-made claim that can be presented in the form of answer, I do believe that an extended dialogue about this can improve the situation. What is important however, is to specify who are to be the partners in such a dialogue and what paradigm of science should be used to lay down its boundary conditions. This is important because unlike the highly industrialized western countries, India is standing at the threshold of a transition, where such a choice would crucially determine its future plan of action.

While talking of transition, it is important to recall the history of modern science in India, which started not as an organic growth but was introduced in the shape of a painful graft. Science originated, or even coincided with, the intense colonial explorations that consolidated the Raj on the subcontinent. Being plunged into this somewhat violent process that was the beginning of science in modern India, it is not unnatural that it often reacts to science today with the lack of an integrated strategy and a divided response. For two hundred years, science assisted a long drawn out process of colonial exploration in the country, keeping a flow of cheap raw materials intact for Britain's industries and providing a market for its finished goods. As a byproduct of this long process, finally a professional middle class was created from the relatively privileged sections of the Indian society, who, from early twentieth century, came to look upon the business of science as their means of livelihood. But the integration of this professional middle class with the rest of the society remains problematic, and often their concerns fail to be in line with the needs of the larger socioeconomic body. Recognition of this divide has led to a kind of sharp reaction in recent times in the form of an Indian science war. Scholars, who, in recent times, delved deep into the nature of colonial engagement,[7] have often concluded that science necessarily violates nature and have espoused the cause of what has recently been dubbed as "ethnoscience," that is, sciences that are local to certain cultures. This was not a move that was observed earlier in the history of Indian science. Paradoxically, we find that the generations, which directly struggled under the colonial yoke and perhaps needed such concepts the most, had not felt need to use them, even though this would have surely made their work easier. The early Indian science professionals like Prof. J. C. Bose or P. C. Ray, clearly wanted Indian achievement and participation in modernity and especially in science, but they wanted it to hold true for the rest of the world and not only be locally applicable to themselves. Yet such early practitioners of Indian science also clearly hoped that a certain kind of resonance exists between India's old tradition and the new knowledge forms, so that the pursuit of science can be aligned with its own classical heritage and can even be seen as its reaffirmation. This process has elsewhere been called the process of moral legitimation.[8] The early Indian names in science took this legitimation as a worthwhile project and even identified this with the greater project of the recovery of the national self-image. A number of prominent cultural interlocutors, from Rammohan Roy[9] to Vivekananda,[10] who did not directly work in sciences but decidedly shaped the modern Indian culture as it exists today, put unprecedented emphasis on this, believing that India can be truly modernized only when this process is brought into completion.

Which Image of Science?
Exploring between the Old and the New

But what image of science are we talking about, and what is its source? It is sensible to ask now what the term *science* stands for before we move on to ask the second question if such a science can be aligned with Indian classical traditions in the project of its development and modernization. What image of science was being referred to when we identified as "scientific" the new bodies of knowledge that arrived in India at the beginning of the nineteenth century? It is necessary to explore the dominant conception of science in order to understand this question that consciously or unconsciously formed such discussions for the better part of a century and is only now beginning to be replaced in the light of historical evidences and social consequences. What science is or ought to be seems to have been well understood in the beginning of the last century. But this situation has undergone some deep changes in the last forty years, following the rise of the Kuhnian picture of science as a succession of non-cumulative paradigms interspersed between periods of a normal, conservative activity. The unity-of-science's program has lost much of its force as a result of this conclusion, which seems to be historically well supported. With the decline of the unity-of-science program, rethinking about the explanation of science has begun. The vast majority still agrees that science works and is probably the best thing that has ever worked, but a general understanding of why it works seems to be remarkably unclear. The different accounts of explanation of the production of scientific knowledge look at different mechanisms to explain why science is reliable as well as powerful, but the emerging consensus is that such explanations cannot be given in terms of a priori methodology. There is no master narrative of science that will be eventually discovered in every culture. From the unity-of-science program, we are moving towards a picture of disunity and family resemblance.

A word about the historical circumstances under which science was born may be useful here. Science grew up in its European homelands in direct confrontation with religion, or at least with the Christian church, and succeeded in gaining a foothold in the imaginations of men. This growth occurred not only because it delivered certain answers about the natural processes but also because it offered a new kind of worldview that promised to liberate people from their existing limitations. Modernity had definite expectations from science and its attendant worldview, not only as a source of possible answers but also as a body of thought that would eventually guide which questions are worthwhile to ask. The questions science would not be answering are not questions about the nature of things. Such questions might arise, for instance, in aesthetics or morality. But in so far as we are to find our way around the world, science is the only method for answering such questions.

The paradigm that today we know as science emerged over the seventeenth and eighteenth centuries and a new, competing tradition was born to replace the religious narrative, gathering around itself a new kind of hope. A continuous worldview was formed as a result of the cumulative work of Copernicus, Galileo, Newton, and many others, and gradually this new tradition came to be pitted against the older, religious worldview. It offered to explain whatever the older tradition explained, but by a set of different mechanisms. The place of God was taken over by the concept of unification or laws of nature. Both traditions shared this one characteristic however, that they, given enough time, claimed to emerge as master narratives, offering to connect everything in a single scheme of things. Both were in a sense jealous traditions, that is, they did not have room for any other competing schemes, any other world views.

Logical Positivists in the 1920s formulated this intuitive expectation into a rigid requirement. In an essay written in the 1950s, Feigl formulated the austere, classical outlines of a science whose aim was to cover a maximum number of facts by a minimum number of laws.[11] Visualized in this way, science is not responsible for the changing landscapes or dilemmas of developments. Morally at least, science has little to do with the technological consequences that arise from it; it is not marked by its successes and failures. The consequences of science do not belong to it, which is contextless. It is this ideal of science that was exported to different non-western parts of the world and was enthusiastically adopted by their local elite populations, who saw in it a cure for all their local maladies. Science was, for a while, the only worldview in conformity with the order of things that would, eventually, supersede everything else. Let me call this, for the sake of argument, the utopian view of science. It is precisely this utopian image that is responsible for dilemmas of development today, which arise when the local and long-term consequences are ignored in the hope that science can eventually fix everything. With the passage of time when every society on earth is becoming more and more deeply engaged with science, it is becoming increasingly clear that this older concept needs to be examined. It stands upon a number of assumptions that can be upset easily with historical and sociological critiques.

Utopian science, to urge the point once again, was wedded to a unity-of-science program that saw science as a master narrative, which can remain totally separate from the contexts in which it is practiced and produced. Because science was primarily viewed as an epistemological machine, this analysis overlooked the economic fact that science demands resources to be pursued within a certain society and hence is always generated within a given social and cultural context. This point was not so obvious in the days of small and inexpensive science. But in the Big Science of today,[12] it is almost self-evident that science requires enough resources to be pursued and cannot be considered wholly innocent of the consequences that are possible through it. Given the economics of science and how science in practice works, it seems that we are not going to expand our knowledge infinitely in all directions (as Feigl fondly hoped) in virtue of a method but only in some chosen areas, which are patronized by a market. That means in real-life science there will be less explored areas between patches of clearer knowledge, giving rise to a bigger interface with ignorance.

A consumer-driven market that wishes to pay for certain kinds of research, but not for others, will naturally lead to a proliferation of technologies with more or less ambiguous consequences. This means a coming era of very different kinds of technologies that will bring about an era of increased self-doubt and uncertainty, often with an increased power to harm. Yet every society that engages science to solve its problems will have to make some inevitable decisions. Such decisions will include policies about what kind of research a society should opt for what are the probable consequences of such research, and what are the best ways to utilize natural resources. Our knowledge will surely increase as time goes by but often it may not be enough to wisely guide our abilities. Indeed our ability to do things in terms of technology has already greatly outstripped the little we know, and as the future indicates, such abilities are going to multiply. Scientific practice is giving us new goods every day in the form of increasingly powerful technologies that increase our options. But the total picture of where we are going with those options is not often well understood. The technologies alone do not tell us what we are doing to others or to the environment or to our future generations. The context of science can no longer be left out of science but must be counted as part of its internal machinery. The earlier, overly epistemic picture of science has introduced a mindset where a society expects to be completely happy when it can rely on the advice of a few experts, who are in charge of a method and can act as the high priests of knowledge. This hope is proving to be hard to sustain about the future of science as it is evolving today. The responsibility for developing and using technologies has to be extended to include the market of science with its extended and, often hidden, network of actors who influence the making of science. Science can no longer be understood in terms of a methodology, or even as a body of expert judgments, but must be seen as a mode of social production of knowledge, demanding the use of social categories. This wider sociological process cannot be left out of the epistemological considerations by simply naming them as irrelevant contexts of discovery.

The danger of a purely epistemic, contextless science has recently been felt most sharply in biomedicine where human stem cell research and other possible genetic manipulations are ushering in a new kind of medical practice. Regenerative medicine opens up new possibilities both in medicine and in research but also raises the specter of possible scientific horrors like genetic prospecting, harvesting of biological materials (which seems to have entered the list of newly possible kinds of crime), and the possible creation of a genetic underclass. Already it has raised questions about the identity and dignity of persons, the nature of informed consent, an over all concern for the eco-system, and what it is that we have to ultimately respect. Responses to such technologies are now increasingly being sought within other alternative thought systems available to the western world such as the Catholic, Protestant, Jewish, or feminist worldviews. Curiously, the role they are playing this time is a scientific one, by staging a dialogue between science and other forces within the society of its origin. They are helping to raise questions that, for a change, are coming from science. Perhaps the utopian picture of science as the single body of knowledge capable of answering all questions (given enough time) was not historically wise. There are questions after all where science alone has no plausible answers and they are more than questions about emotional attitudes. The future seems to be moving more towards a scenario where a pluralistic dialogue between several kinds of knowledge systems may be of more use than the victory of a single one.

The recent accounts of science in terms of its social or cognitive mechanisms highlight the futility of looking for the one unified account of science by showing its widespread social roots and its cognitive continuity with other human tasks. The social normative accounts of science accept that science indeed embodies social negotiations that result from different points of interest in society but such interests are balanced against one another because they all go through an extended social process before anything is absorbed as part of the knowledge system.[13] Instead of isolating a privileged mode of individual reasoning through which everything else is justified and finally grounded, the social-normative account finds science in the very interest-bound forms of social life or, rather, in the wide spread social dialogues that one part of the fabric carries on with another. There are no constitutive values in the sense of a priori method, but scientific knowledge is still possible as long as the channels of communication remain open.

The naturalistic cognitive accounts, similarly, set aside the view of science as justificatory discourse. Science is explained not as an epistemological machine, but as a continuation of our evolutionary cognitive abilities that have grown enormously sophisticated at some point of human history. This evolution has occurred because our cognitive abilities have interacted with external representative structures, making use of language as well as that of a social network to frame this process.[14] This, in brief, is Giere's answer to his own 1492 question about the origins of scientific knowledge. Much of this process is clearly distributed as well as social and takes place outside rather than in the privacy of a human mind. Giving an account of science in terms of only its epistemology is like producing a map of a large territory that can only claim to be perspectival and in a way, always less than complete. There seems to be simply no one scientific way of thinking but several ways of thinking that are all, more or less effectively, goal-directed and can be of equal use while doing science. This is the net cognitive output of the utopian image of science that hoped to demarcate a privileged body of thinking, in comparison to which, all other bodies of thought would to be labeled as prescientific. When we see things in this new naturalistic way, we also begin to see why science might need to make use of other modes of explanations or dialogues, rather than take over all their jobs single-handed. Indeed, the tasks of one can be reformulated and occasionally reexamined by using the language of the other (assemblages of knowledge). And, several ways of knowing can cooperate rather than compete in helping to choose the individual or the societal goals that are sustainable and effective, thereby leading to a worthwhile way of life.

From the older idea of a utopian science that did not recognize any context or culture (because it hoped to transcend them all, given enough time), we are led to the more realistic concept of a science that is sensitive to a context. This "realistic" science functions within a network of social categories and relationships as well as within an eco-system. The function of such a science is to enhance the network it operates within or at least leave it as unharmed as possible. Such a science might often engage in a dialogue with all other knowledge forms that are present around it so that it can preserve and protect what they have achieved. Professional science does not become impossible according to this picture, but perhaps grows more sensitive to its own possible impact on the entire network within which it is practiced. It considers the sociocultural contexts while offering its remedies and sometimes will issue warnings when its remedies are accompanied by grave problems. Instead of believing in laissez faire science, which is expected to solve all problems because of its privileged epistemology, it is time to recognize that, like any other system, science operates within a wider network. An individual or even a small group often has the power to affect or enhance the whole system.

Sciences, then, let us agree do not carry their regulative ideals within themselves as Feigl thought they would but receive them from elsewhere. Precisely for this reason, they require criticism, dialogue, and other similar processes to reach to the level of an informed consensus. More importantly perhaps, they need to be situated with other local bodies of knowledge that contain valuable information but perhaps do not know (or have forgotten) why they are valuable. It is a wise policy for a society to keep alive all its indigenous bodies of thoughts and practices, so that a dialogue, if necessary, can be staged among them. Before giving itself up into the hands of an all-enveloping professional science, a society needs to raise important questions about what this science is going to achieve locally and whether its ways compare with those of others that have stood the test of time. The overriding reason for cultivating science and its close companion, technology, is not because it produces wisdom or enlightenment on its own but because it removes powerful obstacles, thereby leaving human beings free to pursue happiness. But the shape of a happy society needs to be contextualized and cannot be all summed up within a privileged, single discourse.

Where will science finally lead us? To this oft-voiced question in the earlier decades there was a clear and straightforward reply: to Enlightenment. It was hoped that such enlightenment alone would hold enough power to replace all pre-existing bodies of knowledge (which will all be eventually abandoned by the arrival of science). The force of this rhetoric is weakening progressively in its own western home ground. The socioeconomic analysis of science clearly brings out that sciences do not only seek enlightenment, but also seek more mundane things like power, and power often gets concentrated in the hands of particular people or particular groups rather than residing with humanity at large. On a closer look, science involves a network of hidden actors whose roles were not well understood before but cannot be ignored any more, that is, if we wish to understand the process of science adequately. Several cases of selective and biased research in bio-medical sciences have demonstrated that medical research needs some guidelines in the form of risk analysis and other similar protocols in order to prevent doing harm.[15] An account of science given solely in terms of its epistemic mechanisms cannot explain why protocols like risk management and risk communication are increasingly becoming part of the daily routine in bio-medical sciences. Because science produces power but often not enough wisdom about how to utilize that power, it requires clarification and explanation in terms of other cognitive resources of the society and is best perceived as a powerful instrument that requires illumination through a dialogue rather than the other way around. Science cannot be expected to live an uncontaminated life in the heads of a few experts who will always do what is right in virtue of a rational method but needs participation from its greater sociocultural base and from its other local bodies of information. The need of the hour is not only to have science but responsible sciences that are transparent about their own inner workings. This is truer for societies that seek their development through science today because they have historically spent less time with it and may be less aware of its pitfalls.

As an example of how a context-sensitive science might operate in a developing society, let me draw a contrast between two models of agriculture that were present in the history of Indian agricultural development to the eventual victory of one over the other. The concept of modern organic farming, which is now being enthusiastically promoted in the United States and in other industrialized countries, originated apparently in India, not due to an Indian scientist but to an agronomist in the service of the imperial government. Sir Albert Howard[16] and his fellow-scientist and spouse, Gabrielle Howard, lived in India from 1905 to 1930. They headed first the Imperial Agricultural Research Institute at Pusa (1905-1925) and later founded their own institute of Plant Industry at Indore. It was here that they developed the basics of what is nowadays known as the modern organic farming that relies on the principle of manuring the land by using organic wastes, which enhance soil-texture by water retention and aeration and keep the land fertile for longer periods of time. This is in direct contrast with the widely used methods of synthetic fertilization that gives a high output for a short time but leaves the soil increasingly alkaline and exhausted. Tracing the history of Howard's development as a scientist, we find, remarkably, that he was not always in favor of such practices. Indeed, in his early days at the sugar cane plantations in Barbados (sugar being their most important cash crop), Howard had espoused a very different model of agriculture. Organic waste was then for him no more than a source of potential disease and fungus that ought to be promptly destroyed by burning. Reality, according to "early" Howard, does not reveal itself to the peasant in the field, who is likely to do everything wrong. The peasant is required to listen to the scientist and can expect good results only when he is informed by their professional knowledge. Only inside the lab and under the microscope, does reality reveals itself. Agriculture does not need local knowledge.

Remarkably, in his three-decade-long engagement with Indian agriculture, Howard gradually revised his earlier opinion and finally moved towards a hybridized position where he saw both local and professional bodies of knowledge informing and enhancing each other. He noticed that plant and animal diseases were more prevalent on the chemically farmed Government lands than on those of the local farmers who were too poor to buy the chemical fertilizers. The poorer farmers rotated their crops, thus keeping the soil productive and were reluctant to cultivate crops that destroy the soil and produce alkaline wastes but fetch more money on the market. In the practices of these poor native cultivators, Howard found the outlines of a different kind of science where the arrow of knowledge moves from the peasant to the scientist and the farmers act as teachers to the scientific professionals. It is their hard-won bits of wisdom that tell the scientist what has stood the test of time. Yet, in spite of noting how wisdom can be extracted from the face of poverty, Howard did not endorse the currently popular position that science always violates nature. On the other hand, he had visions of a science that informs and is informed by local knowledge. While the farmers had certain instinctive knowledge of what will or will not stand in practice, they certainly did not get everything right. Most importantly, often they could not prevent the waste of their meager resources because they did not know the principles of their practice even when the practices themselves were sound enough. Not knowing the principles of their practice also meant that it was difficult for them to improve upon their old ways, without undergoing innumerable rounds of trial and error. A context-sensitive science that accepts such culturally embedded expertise as valuable can help to show the farmer the big picture of what he is doing and help him to devise better ways. With the valorization of such indigenous knowledge, the scientist accepts the position that certain kinds of agriculture may be customized for certain social and cultural situations and cannot be replaced without causing great imbalance to the ecosystem. What was based on centuries of experience and a native feel for things can be given a better home and better defense and can be practiced with higher productivity at the international market.

It is a point to ponder that these remarkable principles, which Howard discovered on the basis of his teams' long experience in India, were barely mentioned or even used, when India decided to become self-sufficient in food production through the Green Revolution in the 1950s. The dominant technology was largely imported, borrowed from the highly successful methods practiced in the centrally mechanized model that is applied on the Northern American plains. The Government-funded research institutes imported high-yield varieties of wheat from the United States and Mexico and worked upon them to change their appearance for the Indian consumers and make them suitable to the new climatic conditions. Farmers were assisted with fertilizers and other economic packages so that they would find it profitable to plant the new varieties of crops. The result was an apparently grand success story and a tripling of yield that made India self-sufficient in food grains. But at the same time it changed certain things forever. The revolution stamped out many of the locally available best strains because they were not suitable for the homogenized, fertilizer-dependent farming that produced an increasingly saline, waterlogged soil and, more often than not, resulted in a great disruption of the social networks. This led to the curious situation that while food production increased and more food became available in the open market, food entitlement paradoxically failed to keep pace with it—many poor families remained in hunger despite the plenty that surrounded them.

Science as a Colonial Graft: The Indian Case Study

To understand the mindset that accepted the Green Revolution and other similar developments as inevitable, it is important to look at the historical circumstances under which science was born in India. The transformation of an older culture into the modern bilingual society of today was a complicated process at its very best. The engagement with modernity began around 1813 and was officially held to be complete by 1914 when the Indian Science Congress held its first-ever official session. The very mode of that introduction gave birth to a divided mindset among the scientific professionals who were, in a sense, the products of that engagement. In brief, the nature of that historic engagement fostered the general belief that science can only be pursued in the Indian context by being heavily dependent on outside support for both its practical and moral needs. The history of Indian science often reads as the history of a stressful endeavor that continuously oriented itself toward external forces to decide upon its plans and projects, and often make its contacts and earn its laurels. It is this atmosphere of an intellectual as well as emotional dependence on the outside that gives rise to the mindset that science can only travel from the experts, preferably imported, to the peasants in the field, thereby creating episodes of an imbalanced development. In what follows, I shall avoid an in-depth analysis of the process of colonial engagement because that has been provided elsewhere in much greater detail.[17] The highlights of the process explains why Indian science professionals often practice science in the same spirit as their erstwhile empire builders, that is, divorced from their contexts and surroundings. This relationship needs to be changed if Indian engagement with sciences is to prove more beneficial to the society of its origin.

This image of science as borrowed knowledge can be traced back to the choice that was made in 1813 in favor of the western style of education. In the crucial cultural debate that preceded this decision and introduced science in the Indian subcontinent, prominent Indians like Raja Ram Mohan Roy gave their opinion (1823) in support of western science. But the document that shaped the nature of the Western education and the subsequent education in science was Wood's Dispatch,[18] which practically defined what such an education would mean in the Indian context. The Dispatch is considered to be the single most important document in the history of higher education in India because it was the first declaration on the part of the British Government about the educational policy to be followed for its Indian subjects. Accordingly, in 1857, the first three universities were set up in Calcutta, Bombay, and Madras to function as the vehicles of higher education.[19] The function of such universities, however, had little connection with research or spread of education among people and was confined, disappointingly, to the holding of examinations and conferring of degrees. Research formed no part of the original charter of the Calcutta University for example. The universities functioned simply as reliable machines to produce good civil servants to serve the interests of the empire and saved it the expense of importing such personnel from home.

Between the well-known document called Macaulay's Minutes (1835), with its crippling comments on Indians' abilities, and the subsequent declaration of Wood's Dispatch, the main outlines of the British education system had formed. Macaulay recommended that a race must be brought up through education, who, save in the color of their skins, would be European in all other respects. The aim of such education would be to promote loyalty to the empire and to prepare the Indian subjects for limited responsibilities in the administration. Education was offered neither to spread universal literacy nor to promote research but for limited dissemination of information within tight bureaucratic control.

The clear insufficiencies of the government-sponsored education gave rise to the competing concept of a national education, particularly after politically disastrous events like the Partition of Bengal (1905) by Lord Curzon.[20] Such events highlighted the clearly felt need for the technical, engineering, and medical schools which was not being met by the government-sponsored universities and also created an increasing urgency, side by side, to communicate the knowledge of India's art, philosophy, and literature to the outside world. Attempts were made to establish nationally managed educational institutions with the support of the prominent members of the community, and in 1906 the National Council of Education came into existence as an engineering school, which has grown into the Jadavpur University of Calcutta today. The learning of science seemed to be an important way of registering protest.

The professionalization of science in India thus took shape within a strongly bipolar atmosphere. It swung constantly between the image of science as a borrowed form of knowledge, having no deep roots in the Indian psyche, and a personal response to this image by some outstanding individuals in the form of achievements in science. By and large, the professionalization of science took place within the state-sponsored university system because private institutions simply did not have the necessary funds to support education and research in sciences. But the two kinds of institutions, government and private, often worked side by side, in deep personal interaction, and research in science often drew inspiration from the greater project of the recovery of the national self-image. Personal achievement in science was considered equivalent to patriotism and was even seen as a form of rebellion against foreign rule. Within the first three decades of the twentieth century, a number of research institutions were established that quickened the process. Dr. M. L. Sircar established The Indian Association for Cultivation of Science (1876) in Calcutta on the model of the Royal Institution. The first physicist of India, Professor J. C. Bose, was also one of the first to create a professional space for Indian science outside the university system. He established Basu Vijnana Mandir (literally, the Temple of Science) to give freedom of research—which he sorely lacked in his youth—to the coming generation of Indian researchers. He began his career as the first Indian professor in physics in the Presidency College, Calcutta. He presented his work, both on radio waves and plant physiology, on several occasions to the Royal Society of London, thus giving the Indian professional scientific community, for the first time, a face in the outside world. M. N. Saha founded Saha Institute of Nuclear Physics and worked on thermodynamics and statistical mechanics. Professor S. N. Bose, of the Boson particle, chose to work on theoretical physics and relativity and collaborated with Einstein to produce Bose-Einstein statistics. By 1914, the first Indian Science Congress had been organized, and modern science had sufficiently crystallized to take a definite professional shape in India. In 1930, Indian professional science received its highest recognition when Professor C. V. Raman received a Nobel Prize for his work on the Raman spectra. Within the first three decades of the twentieth century, the second-stage practitioners of modern science in India had managed to establish a number of research institutions and professional bodies and had trained a sufficient number of graduate students to ensure the self-replication of a professional scientific community. Science in India had finally emerged as a serious profession, reliable enough to be pursued as a means of living.

Three stages can be discerned in the process of grafting the sciences onto Indian society.[21] The first is the stage of the self-trained brilliant non-professional, who engages in some kind of translation work between the two knowledge systems but is not formally trained in the discipline where he works. Ramanujan is perhaps the best example of such a self-taught genius, who conducted most of his mathematics in the shape of having visions about his theorems, communicated to him personally by his village deity, the account of which completely baffled his collaborator, Hardy. The second-stage Indian science professionals, like Professors S. N. Bose or P. C. Ray, however, are attached to a university system and operate through the standard mechanisms of publishing, teaching, and presenting to professional bodies. In the history of Indian science, the second-stage professionals look upon their own tradition critically, and yet hope to salvage and revitalize as much as they can to realign it with modern sciences. They wish to explore the question of an overlap between their classical heritage and their modern training and ask if one can be used to understand and illuminate the other. The research problems of the third-stage science, however, deviate from this pattern. Once the professional boundaries set in strongly, the research problems are frequently divorced from their contextual settings and the aims of science are explained in value-neutral, universal terms. The research problems are now chosen by keeping an eye onto the problems that are popular in other centers or in more developed countries. They are not chosen on the basis of the contexts of the people (or the landscape) among whom science is practiced, with the implicitly held assumption that what works in one place will equally well work in another. The dialogue between modernity and a situated tradition is abandoned at this juncture, and science is accepted as an all-pervasive natural discourse. Yet, as we have seen earlier, such a utopian science can give rise to skewed episodes of development, when a society is more than ready to forget all its locally produced knowledge and does not want to dwell too long on uncomfortable, long-term questions. Indian science, since independence, apparently operates at this third level, where questions of development are often decided by taking the top-down approach with the assurance that science is the primary means to solve all kinds of questions. Modernity has merged with an image where science is increasingly seen as capable of fixing anything.

Why science is often perceived as a borrowed body of knowledge that cannot be practiced with success unless extensively supported from outside becomes clearer in the light of this ambivalent piece of Indian history. Since the knowledge systems did arrive from outside in the form of implants, they were, first of all, seen as parts of an alien way of life, both by the rulers who brought them in and also by their subjects. But they were unavoidable and powerful and had to be integrated somehow in the Indian forms of life. By the time the relatively well-off, upper-class Indians had made up their minds to embrace science as a worthy profession, reactions to sciences were dominated by two kinds of responses. They first responded by rejecting them as not *being* in line with what is traditional and hence not participating. (Interestingly, the Muslims in India largely went with this option at least in the beginning.) Or, more importantly, they responded by imitating the successful practitioners and relying heavily on their practices for problem selection and policy making. Even the second-stage science professionals in India, who came nearest to developing a different kind of mindset that is not restricted to either of these constricting attitudes, still relied heavily on the outside support in the form of their erstwhile teachers or friends before they could practice any science at all. The overall picture of working with sciences in the British India was that of an intellectual dependence and nurture from outside, because little nurturing was available at home from a colonial administration, which did not have a great deal of confidence in the local abilities. Naturally, this torturous history left a mark on the subsequent trajectory and problem choice of Indian science, which remained in the grip of a colonial hangover, even after independence, particularly in the formation of its science policies. Most science professionals in India today see the making of scientific knowledge as an enterprise where one is always dependent on the expertise and opinions from outside before any lasting contribution can be made in a specific field. The makers of science in India often see themselves placed on the peripheries of knowledge making, where they constantly need to orient themselves (and their work) towards a certain center, before it can be placed in the world at large. Along with this mindset, there is also the belief, more or less openly expressed, that a great deal of fault lies in the classical Indian traditions, which cannot sustain a scientific attitude among the masses on its own. Several scholars in Indian science studies expressed the opinion that Indian tradition is an inadequate vehicle to support the true scientific attitude and can be of little use in making a new science or an enlightened society.[22] The hallmark of colonialism, limited confidence in local abilities, persists long after the days of colonialism are over.

The Classical Tradition: Case Study Continued

I opened up the discussion by suggesting that an alignment of worldviews is possible between science and the classical ideals of the Indian philosophical traditions. To develop this side of the argument, let me now sketch, in brief, what India had traditionally understood as freedom and as a worthwhile way of life. It is reasonable to ask, at the end of our discussion, what did it have in its traditions and in what sense can those ideals contribute to the making of a context-sensitive science that helps to form the concept of a development without harm?

I shall begin exploring the question by pointing out the most striking characteristic that is always noted in an introductory discussion of Indian philosophical systems, and which is sometimes explained as their practical character. Uniformly, the Indian systems begin not by being concerned with the glory of God but from the most obvious fact of life, the fact of human suffering. Human life, in general, seems to be characterized by certain kinds of suffering that demand to be solved. The task these Darshanas (literally, perspectives or ways of seeing)[23] set for themselves is not to deliberate upon the existence or non-existence of God, or even of an external world, but to solve the problem of human suffering (Dukkha). In a clear Socratic orientation, the Indian traditions begin not from God but from man. Each system therefore begins with the deliberation of all possible kinds of sufferings human life can be subjected to, but also emphasizes that suffering can be eliminated, if only we develop the right set of worldviews and have a program to translate them into action. Suffering arises, according to classical Indian thought, because we labor under ignorance or Avidya (literally, lack of knowledge), and is removed as soon as the individual informs herself about the nature of things. Pain and suffering have no ultimate ontological existence, and the goal of life is to arrive at the blissful realization of our own Self. The nature of Self (Atman)[24] is identical with the universe and only ignorance about the status of the finite individual (jiva) prevents us from being in contact with this realization. The reason why we need a Darshan at all is therapeutic[25]—it is needed in order to neutralize the effects of our own previously misplaced actions, to remember the essential blissfulness of the Self, and to help us develop a program of correct actions. A person, however, needs correct goals before she can decide on the correct actions that will achieve this definite result. And this in turn implies that she needs a worldview within which she can see herself connected to a larger network. Epistemology and metaphysics are thus important aids to the recovery of the Self. A happy life is directed towards goals that are correct. This leads us to the most important core concept of Karma in the Indian tradition, on which all systems converge to emphasize the deep, long-lasting effects of our own actions and their importance.[26] An interesting point to note is that the unit of suffering is always understood to be an individual. The passions and ignorance of the individual, the traditional images tell us, are responsible for keeping our true nature hidden, which results in suffering. The function of philosophically informed actions is to remove this layer of impurity and to neutralize the effect of one's own previous ignorance. Self-aware actions help a person to reach her new goals by literally making a platform on which she can place herself. As the mountaineer slowly climbs upwards by using a rope that is coiled around her waist, so do we start our upward climb towards our own selves by holding onto suitably informed actions infused with a complete worldview.

The right knowledge that dawns at the end of such inquiry makes a person free by placing her in connection with the universe and uniting her with all living beings. The summit of all Karmas and the end of all suffering lies in this conception of Moksha, (literally, being set free). It is an overall awakening of the individual, who had been temporarily enveloped by ignorance by forgetting her own nature. When such nature is recovered, there is an erasure of the boundary between the Self and the other, between what was (formerly) in the inside and the outside. The realm within and the realm outside are seen continuous in the eyes of the person who has regained her freedom. The aim of the spiritual exercise is clearly not to contemplate the glory of God, but to recover the happiness and freedom of man. The texts of the tradition suggest that such recovery can be done thorough work, worship, logical reflection, and performance of ritual actions or by any other thing whatsoever. But the exact path that is to be followed is always open to the choice of the individual.

What is the final shape of the person who has been set free at last? I quote a fragment from the Indian epic *Mahabharata* that gives a quick look at the person who has freed herself from the bound state.

> Unselfish, without egotism, free from the pairs,[27]
> Having cut off doubts, he is not angry and does not hate,
> Nor does he speak false words . . . .
> His sense organs are not drawn to many objects,
> His desires are not widely scattered,
> He is not harmful to any creature,
> Such a Samkhya-follower is released.[28]

A limitation of this traditionally conceived task is perhaps that it did not clearly speak of social or collective suffering or sufferings that arise due to the effects of wider historical circumstances. Suffering is largely understood in these philosophical texts in terms of a solitary individual, and its cure is also similarly prescribed. A possible explanation for this overly complacent attitude was probably India's false sense of security due its well-insulated geographical position. Protected by the great Himalayan chains in the north and by the three oceans in the south, the early Indians must have thought that individual suffering is as far as suffering will ever go. History has proved them sadly wrong. Apparently, the causes of suffering could be, and have been, even more manifold than was ever thought of by our distant ancestors. What we can do today to respond to those lessons taught by history. We are to give an extended interpretation of our bondage, so as to include the wider network of socially shared Dukkhas and to articulate the concept of the overall development of the whole social body, in terms of its material, social, and economic conditions. It is no longer possible to address the problem of suffering or even to eliminate it by thinking purely in individual terms. The possibility of this wider interpretation is not exactly unknown[29] but is not always sufficiently highlighted. This less-than-adequate articulation of the ideal has produced the remarkable effect that Indian tradition is often understood as to be synonymous with a set of discriminations and oppression of one social class by another.

I think the new task departs from the old in one significant respect. The classical way of formulating the problem was to look into the self of the individual to identify the shackles of attachment and ignorance in order to break asunder any bonds that contributed to suffering. In the texts of the tradition, the problem of human life and its eventual solution is always laid out in terms of removing the psychological obstructions that lie within the individual. But such obstructions that we see in today's world need not be construed only as faulty thoughts or emotional processes of the individual. They can just as easily be understood as larger faults present in the wider collective processes such as social or economic situations. Collective concern with such problems can equally well be considered as steps in substantially the same direction (of producing a happy human life). Liberation need not be conceived exclusively in individual terms but could conceivably be given a social dimension.

It is important to notice in this context that Indian systems hardly ever talk of spirituality in the sense of neglecting the concerns of this world.[30] Rather, they spend their time in developing the concepts of happiness and freedom within it. One is free not after death but in life. The goal is simply to become an awakened individual (whatever the means might be), and this is something that does not have to wait for an afterlife. The classical Indian systems spend considerable time developing the concept of an awakened person who has realized that the world is a system of networks and is, in a sense, an extended image of herself. Her freedom makes her responsible and aware of her actions because they leave a deep impact upon the world, which in a sense, is her own self. The awakened person, as Buddhism tells us, behaves without any thought of harm towards any living creature, and sees herself in everything. Freedom does not mean destruction of desires but simply seeing to it that desires are not scattered in all directions.

The image of an awakened person embedded in a network can serve as a metaphysical research program for Indian society today while it engages with science and its inseparable companion, technology. This seems particularly relevant, because when science was grafted onto the Indian culture, it was still expected that science would function as its own regulative ideal and, given enough time, would eventually supersede all other forms of knowledge. But the very concept of science has gone through a sharp change since those days. It is now abundantly clear that science cannot deliver such promised ideals unassisted, particularly when it is divorced from the contexts of its practice. Modern science does not function in virtue of its epistemology alone, but because it is a social as well as an economic process. As such a process, it is closely interwoven with the rest of its social fabric, including the other cognitive products that are present within it. Without participation from those wider social processes and the rest of its resources, the goods of science would perhaps be less beneficial to the society of its origin.

A Modest Conclusion: Continuity of Concerns

It is time now to wrap up the story by giving it a moral. What is the lesson that we can derive from this tangled mass of deliberation? I would like to suggest that here indeed is a moral for us to take home. Put briefly, an Indian engagement with sciences should not take the form of borrowed strategies adopted from elsewhere that are not organically related to its contexts or traditional ways of life. At the same time, those traditions need to be revised and reshaped by giving them a new interpretation that brings them in line with the projects of today. I would like to suggest that India's classical traditions could be useful to shape and assess its modern projects. Indeed, the classical goals like freedom or removal of suffering can even be better addressed today by the new tools of modernity. Such newly extended deliberations would still be in line with those classical projects because they are not fundamentally at variance with the spirit of those those aims, but are complementary to them. The technological tools provided by sciences can even be seen as part of the newly extended "spiritual" program. While science is thus accepted as part of our own form of life, it is necessary to keep in mind the circumstances under which science was introduced in India and the dominant motives behind that introduction. It was introduced, roughly speaking, not with any intention to enlighten but to enforce a greater control on the country's raw materials and human resources. By and large, this has led to the not too happy situation that in the history of Indian sciences much of the local and traditional knowledge was neglected to the point of being lost. Or, it was simply ignored as not being worthy of preservation because it was not in the interest of the empire builders to do so. Even when the days of the Raj were over at last, a large colonial legacy remained behind in the shape of a mindset that believed in quick borrowings from outside rather than looking within for means that are already available, albeit in need of revision or rethinking. It would not be entirely incorrect to say that enhancement or preservation of local knowledge or its systematization did not figure prominently in the projects of post-colonial Indian science, which saw development largely in the image of borrowing and transplanting of technologies. The mode of interaction with sciences thus remained largely colonial in nature, even though at this point of time, as an independent country, India does have the power of choice that it lacked earlier. But as we have seen, development under such an image is an unsafe development at its best, because of its unforeseen and disastrous consequences. A body of reflections on science and its attendant sociocultural underpinnings is necessary in India to replace this legacy and to bring about a mindset that is able to make better decisions on the question of development and its future engagement with sciences. Such a process of moral legitimation, I believe, is useful in the Indian context, because in the discourse of modern technoscience, the questions that we ask about nature are often becoming indistinguishable from questions about ourselves. The sciences that a society cultivates represents the sum total of human choices made within that society, and this, to some extent, means that we need to first train our choices before we exercise the option of bringing them into existence.

The process of moral legitimation formed an important part of the projects of early Indian scientists like J. C. Bose, who drew inspiration for their science out of possible areas of resonance between tradition and modernity. At some point, when the professional science became firmly established and it was no longer necessary to cushion oneself from the brunt of colonial experiences, such efforts were discontinued. They can, however, be reused today with a very different aim in mind. I want to propose the modest thesis that such older projects are not necessarily erased or suspended by modernity and can be seen as a continuation of the same kind of efforts.

Notes

1. The most familiar Indian prayer from the Upanishads, 1000 B.C. See Internet Indian History Sourcebook in "The Formation of Religious Traditions," http://150.108.2.20/hasal/index.html, accessed April 15, 2002, ed., Paul Halsall (halsall@murray.fordham.edu).

2. In what follows I would like to thank Casey N. McGinnis for his suggestions to improve my English.

3. Alison Gopnik, "Scientist as Child," *Philosophy of Science* 63, no 4 (Dec. 1996), 484-514. Giere's 1492 question brings home the fact that the human cognitive apparatus has remained much the same between 1492 and today. Yet, apparently, science has not. So, what makes the difference between 1492 and today?

4. Tagore, in conversation with H. G. Wells (Geneva, 1930) expressed the opinion that Japan could make rapid strides into modernity because, unlike India, it could absorb and reject the western scientific heritage from its own point of view. It is interesting to note that Japan spends huge amounts of money in basic research today, its budget being comparable to other industrialized nations of the West.

5. *Raj* is a Hindi word that in the history of colonial India refers to the British Empire.

6. Meera Nanda, for example, holds such a view. See Nanda, "The Epistemic Charity of the Social Constructivist Critics of Science and Why the Third World Should Refuse the Offer" in A *House Built on Sand*, ed. Noretta Koertge (New York, Oxford: Oxford University Press, 1998), 286-311.

7. For the stages of displacement and re-negotiation of the colonial encounter, see Dhruv Raina, "Beyond the Diffusionist History of Colonial Science," *Social Epistemol-*

ogy 12, no. 2 (1998), 203-13. Also see Dhruv Raina, "Reconfiguring the Centre: The Structure of Scientific Exchanges Between Colonial India and Europe," *Minerva* 34, (1996), 161-76.

8. See Dhruv Raina and Irfan Habib, "The Moral Legitimation of Modern Science: Bhadrolok Reflections on Theories of Evolution," *Social Studies of Science* 26 (1996), 9-42.

9. Humanist and religious reformer, Raja Ram Mohan Roy (1772-1833), was considered by Tagore the first modern man India produced. He was the founder of the religious movement called Brahmo Samaj and was also India's first feminist who carried on a long crusade to outlaw the practice of self-immolation of Hindu widows, which was subsequently outlawed in 1829.

10. Swami Vivekananda (1863-1902) founded the Ramkrishna Order in 1898 and interpreted India's religion and culture to the West. He leapt into fame suddenly at the Parliament of Religions in Chicago in 1893. Many Indian political leaders have acknowledged their indebtedness to Vivekananda's vision of modern India.

11. See Herbert Feigl, *Inquiries and Provocations, Selected writings, 1929-1974*, ed. R. S. Cohen, (Dordrecht: D. Reidel Publishing Company, 1981).

12. Anthropologists of science like Sharon Traweek define Big Science by the following characteristics: first, it is extremely expensive, the budget of a basic research lab running into millions of dollars; secondly, access to the extremely expensive equipment remains strictly limited, often decided through a panel of experts; and third, because of such restrictions, the questions that are to be investigated have to be decided years ahead. There is a convergence among the scientists about which questions are or are not worthy of investigation.

13. Helen Longino, *Science as Social Knowledge* (Princeton, N. J.: Princeton University Press, 1990).

14. Ronald Giere, *Science without Laws* (Chicago: The University of Chicago Press, 1999).

15. Most of the U.S. institutions and universities that today receive federal grants for biomedical research are subjected to Institutional Review Boards. The Swedish Council in Europe laid out a four-point document in 1982 for similar research, defining informed consent and the guidelines for future use for the data produced in research, thereby underlining the social need for a wider deliberation on what was previously considered to be the domain of a few experts.

16. See Thomas F. Gieryn, *Cultural Boundaries of Science*, Chapter V (Chicago: The University of Chicago Press, 1999).

17. See Dhruv Raina, "The Present in the Past: Trajectories for the Social History of Science," in *India: Another Millennium?* ed. Romila Thapar (New Delhi: Viking, 2000).

18. See Internet Indian History sourcebook in "Western Intrusion, British Rule," http://150.108.2.20/halsal/index.html, accessed April 15, 2002, ed., Paul Halsall (halsall@murray.fordham.edu). Sir Charles Wood, the then Secretary of the State, sent a dispatch (popularly known as the Wood's Dispatch) to the Board of Directors of the East India Company in 1854. He sent the dispatch because by 1853 a number of problems concerning education had arisen in the country that called for a clear act of policy. Wood's Dispatch forms a landmark in the history of higher education in India. In spite of its apparent good intention however, the Dispatch failed to build an interactive system of education that fosters research or encourages originality in the learners. In effect, it proved more useful to the government and to the missionaries rather than to the people. The government got an assured supply of locally trained civil servants and the missionaries could continue their religious work supported by the grants-in-aid provided to the

institutions. The Dispatch did not recognize any obligation of the State to educate every child under a certain age. Neither did it allow for industrial regeneration or any other kinds of intellectual leadership. The spread of education among the people was supposed to take place by natural percolation alone, which at its best, is a very doubtful mechanism.

19. But mostly they remained confined to the administrative headquarters and port cities.

20. Even though Lord George Nathaniel Curzon, Viceroy of India (1898-1905) and British Foreign Secretary (1919-24), was well known for his other accomplishments, he is mainly remembered as the man who partitioned Bengal and unleashed the first mass movement against it.

21. Dhruv Riana, "The Young P. C. Ray and the Inauguration of the Social History of Science in India," *Science, Technology and Society* 2, no 1 (1997), 1-40.

22. See Meera Nanda, "A Broken People Defend Science: Reconstructing the Deweyan Buddha of Indian Dalits," *Social Epistemology* 15, no. 4 (2001), 335-365. Nanda argues that Indian tradition can never possibly be a vehicle of true scientific attitude because it cannot lend itself to criticisms and is wedded to fixed metaphysical doctrines like rebirth and Law of Karma. She interprets Karma as a kind of fatalism which preaches that people are supposed to pay for their past sins by undergoing sufferings in the present life. It is doubtful if this interpretation could be borne out by an examination of the traditional main texts.

23. Philosophical systems tracing their descent from the Vedas, the earliest texts in India.

24. Atman is used as a technical term in Indian philosophy to mean Self but is distinct from the Mind, which is material and is simply considered a master organ in the hands of the Self. Atman is the unmanifested knower of the knowledge field, a limit of the world, in an almost Wittgensteinian sense.

25. Buddhism for example, uses medical metaphors to explain its tenets: suffering (Dukkha), cause of suffering (Samudaya), elimination of the suffering (Nirodha), and a program providing its cure and prevention (Marga).

26. Law of Karma simply means that the result of one's actions endures in the sense of a permanent trace within the character of the individual, till they are exhausted either in their due course or are neutralized by some counter actions. What the law seems to forbid is that freedom could be obtained through grace or any other kind of direct divine interventions.

27. Like love and hate, desire and aversion and all other things, which come in binaries.

28. See Internet Indian History Sourcebook in "The Formation of Religious Traditions." The Epic *Mahabharata*, Book XII.

29. Buddhism, for example, concerned itself deeply with social as well as individual problems, being the first missionary religion of the world.

30. The word for religion is *Dharma*, even though the English word is routinely misleading. Dharma refers more to the natural *excellence* of the individual and has no inbuilt associations to *a* church or god.

Chapter 8

Pirates and the Geography of Knowledge: America and Algiers in the Late Eighteenth Century

Lawrence A. Peskin

There is a spatial dimension to the acquisition of knowledge, just as there is a spatial dimension to trade, the acquisition of material goods. This geography of knowledge may be very limited: for example in an early South Sea island where no information comes from beyond the immediate community. At the opposite extreme, one might easily imagine a rather antisocial New Yorker locked in a windowless office and logged on to an Australian chat room so that his only source of outside information comes from the other side of the earth. There is also a temporal dimension to these two imaginary examples: the second is clearly more "modern" than the first. The geography of knowledge may not expand in every era, but it certainly has tended to do so over the long run. In this regard, knowledge acts like any other commodity: as human contacts expand it diffuses ever further from its source.

As the crucible of both modern capitalism and modern science, the eighteenth century was a crucial period for the diffusion of knowledge. During this period a myriad of agents ranging from intimate salons, to philosophical societies, to correspondence networks, to widely-circulated printed publications accelerated the diffusion of knowledge. Contemporary developments such as the establishment of liberal states, expanding trade, and the growth of printing all helped to facilitate what is still termed the Enlightenment. Despite the high-culture implications of this label, the process was not limited to philosophers and statesmen. Knowledge diffused at all levels of society, rich and poor, urban and rural, to create what has sometimes been called a "village enlightenment" alongside the more glamorous and well-known world of the *philosophes*.[1] Since the turn toward history from below in the 1960s these popular enlightenments have been of great interest to historians. Studies of mechanic institutes, sailors' information networks, women's reading, book circulation, and many other topics have broadened our understanding of eighteenth-century information diffusion.[2] Many of these studies have benefitted from Jurgen Habermas' seminal work, *The Structural Transformation of the Public Sphere*, particularly in its conception of the public sphere as a developing arena of free, or relatively free, information exchange that blossomed in the eighteenth century with the rise of liberal governments, growing world trade, and a powerful bourgeoisie. While they might argue that Habermas situated the public sphere too narrowly in the world of the bourgeoisie or that he ignored the role of class and political contestations in shaping the public sphere, historians nonetheless find the idea of a "public sphere" useful in providing an apparatus to conceptualize information diffusion beyond the immediate control of the state and beneath the lofty intellectual and social heights of the *philosophes*.[3]

The project outlined in this essay is essentially a popular epistemology using a single set of events as a case study to determine the source and extent of public understanding of foreign affairs. It creates a tentative geography of knowledge, a rough map of one small portion of the late-eighteenth century public sphere. Many fine projects, such as the Jefferson Papers or the Hartlib papers have enabled historians and philosophers to map the intellectual world of the *philosophes*. It is much harder to trace the geography of popular knowledge, but that does not mean that it did not exist or was not expanding. It is still possible for the historian to find clues, remnants of conversations, letters, or printed communication that broadened the intellectual world of the eighteenth-century masses. Such a project prompts two further questions: what developments led to this expanding mental map, and what affect did it have on members of the public?

The specific case discussed here is an unusual one. In 1785 Algerian pirates captured two American ships and their crews in the Mediterranean, and in 1793 they captured eleven more, holding a total of approximately 120 American sailors in "slavery," some for over a decade. This was the new nation's first foreign policy crisis, and the emerging mass media kept Americans of all sorts surprisingly well informed of its details. At the start of the crisis newspapers printed descriptions of the Barbary Coast states, and as it continued one of the nation's most prominent publishers put together a history of Algiers as a means of feeding the public's curiosity about this distant enemy.[4] Although the location was exotic, the public appetite for foreign affairs was not atypical. There can be no doubt that the average late-eighteenth-century reader was better informed on foreign affairs than readers of any previous era and of most succeeding eras, too. Many American newspapers contained virtually no local news, but all contained at least several columns of reports from abroad invariably featured more prominently than local occurrences.[5] Most Americans still lived within a few miles of the ocean and were closely connected to the Atlantic world by the ships that daily entered American harbors with European and African products and information and left with American produce and sailors. Their world was far from parochial. Yet, from our perspective, their knowledge of it was limited by the speed of communications, which could not travel faster than the fleetest ship. Scrutiny of the reaction to the Algerian captures offers insight into how individuals were affected by globalization at the same time that it suggests how a widening geography of knowledge may have helped to shape the nation as a whole.

The Technology of Knowledge, 1785-1792

Although the American captives suffered thousands of miles away in a place virtually none of their countrymen had visited, their fate made a vivid impression on all levels of society. Louis Guillaume Otto, the French Consul in New York wrote Thomas Jefferson, "The hostilities of the Barbarian corsairs have made a great sensation in America." Jefferson himself, although publicly calm, was privately terrified that his young daughters might become the next victims when they crossed the Atlantic to join him in Paris. A widely reported rumor held that Benjamin Franklin was captured as he returned home from France. Nervous Virginians imprisoned and deported an unfortunate pair of Algerians who soon reappeared in Charleston, where they were once again imprisoned and deported. In New Jersey a sailor and grifter calling himself Charles Blinckhorn attempted to extort money from Governor William Livingston by claiming that he had found his missing son imprisoned in Algiers.[6] The latter is a particularly interesting example of how a man on the bottom rungs of society was not merely aware that Americans had been captured in Algiers but was knowledgeable enough of the details to craft a story that nearly fooled as sophisticated an individual as William Livingston.

Much of the information technology behind this widespread awareness was no more advanced than the sailing ship. There were four basic components to late eighteenth-century news gathering: oral communications, letters, official documents, and newspapers, all of which were carried by ships. The first three were not new at all. Only the fourth, newspapers and other printed material, could be considered at all cutting edge at this time.

The first messenger to inform Americans of the Algerian captures provided an oral report. Captain McComb of the Massachusetts ship *Rambler* happened to be docked in Cadiz harbor when word of the captures reached that Spanish Atlantic port. McComb and his crew had nearly become captives themselves a few days earlier on their approach to Cadiz. The crews of two Algerian galliots blocked their path and began to quiz McComb on his ship's origins and cargo. Aware that Americans were being targeted, McComb lied that his ship originated in Ireland, and he displayed a British flag that he apparently kept on hand for such emergencies. His return trip was smoother, no doubt because of the Portuguese convoy that now protected the *Rambler* and other ships from the North Africans. On arrival in Beverly, Massachusetts, fifty-four days later he would rush to share his information with the Americans who met him on shore.[7]

Historians' reliance on written sources can easily blind us to the continuing importance of oral communications such as McComb's. But, as the Algerian captures show, oral reports were crucially important in an eighteenth-century world of low literacy and limited print media. In general, oral communications from Algiers fit into two broad categories: the vast majority came from ship captains such as McComb and their crews reporting on possible "barbarian" captures of American vessels, while a second, smaller group consisted of reports from visiting strangers like Blinckhorn claiming to have been held in captivity themselves. The reports of ship captures were of vital importance to mercantile towns where a very large proportion of the population would be directly affected by such events. Merchants stood to lose their fortunes. Seafarers' wives and children could lose not only their loved ones but also their primary means of support. All of the shipping-related artisanal trades—shipbuilders, riggers, caulkers, chandlers, sailmakers, blacksmiths, and many others—stood to lose a portion of their livelihoods. And should the Algerians become too active they could potentially shut down America's Mediterranean trade, thereby seriously impairing the entire economy.

To provide vital information to an anxious public, the seafaring world created a long distance version of the children's game of telephone. Consequently, there was an element of randomness to oral reports, for they depended on the timely arrival of somebody who happened to have new information. The newsbearer sometimes witnessed events himself or herself.[8] More frequently, however, reports came second-, third-, or even fourth-hand. As in the game of telephone, the reports furthest removed from the original source tended to be more distorted and more sensational than firsthand reports. For example, a Captain Gage arrived in Boston after a ninety-day journey from Lisbon to report that he had encountered a Captain Cunningham of Petersburgh who reported that an American brig bound from Petersburgh to Lisbon had been captured by Algerians. Captain Gage brought information about as rapidly as possible, that is, in roughly the time it took a ship to move from the site of the event to the port city waiting for news. However, his report, like many similar oral reports, proved false. Somewhere along the line the information had been distorted.[9]

The second group of oral reports, accounts of captivity and escape from the Algerians, were even less reliable than those from the seafarers' news network. They reveal a group of traveling grifters like Blinckhorn who fashioned the news of the day into confidence scams, the ancestors of the sort of Mark Twain characters whom Huck Finn called "low down humbugs and frauds." Thomas Jefferson encountered one of these men in Paris. He claimed to be Thomas Smith, an American captured by Algerians near Lisbon and held hostage aboard an Algerian ship for more than two years. He said he escaped in Cartagena, but "being entirely destitute of money, and likewise of clothing, too, in some measure," he begged the American minister for money. Jefferson was not fooled by this ploy. For one thing, Smith "could not speak English but very broken." The "captive" tried to finesse this issue first by claiming to be from Charleston, and, when Jefferson remained skeptical, by claiming that he was actually a French Canadian. After Jefferson observed that Smith was hardly a French name, the "captive" decided that he was instead of German extraction. He tried to recover from this line of questioning by displaying a pass, allegedly from the English consul at Cartagena. This document, Jefferson noticed, was "in broken English," and in the same handwriting as the note Smith had earlier written to gain an audience with Jefferson.[10] Neither Smith nor Blinckhorn appear to have been particularly literate. No doubt they gained most of their information about events in Algiers through the seafarers' word-of-mouth grapevine. While both their scams backfired, the fact that they could even make the attempt reveals both the extent of news literally floating around the world and the uncertainties that remained despite the large volume of information.

After oral reports, letters were probably the most common news medium of the day. It is no accident that the earliest newspapers were called "newsletters," for they extended a long tradition of exchanging letters filled with information about world markets and events. Most letters were written by merchants, who as shipowners, had both the means and the motive to keep their overseas correspondents well informed. Merchants generally believed the news in these letters to be more trustworthy either than oral information or than newspapers, and they continued to be a vital source of information long after the advent of printing.[11] In 1785, for example, Captain McComb did not merely bring oral information, he also carried a letter from a Cadiz merchant to a Massachusetts acquaintance detailing the news of the captures.

Letters probably offered a bit more reliability than word of mouth information. Nonetheless, it is still important to remember that the divide between written and oral reports was far from absolute. Letters flowed through the same seafarers' network that carried oral reports back and forth across the Atlantic. The seaports were so hungry for any and all news about shipping and possible impediments to trade that recipients probably did not bother to make fine distinctions between written news and word of mouth. For example, seafarers might orally report the contents of letters they had read earlier when landing at a new port, where others might rewrite the news into letters sent to interested correspondents.[12] But, unlike oral reports, writing suggested a fairly simple method of verification; after all, one of the tests that tripped up Twain's grifters in *Huckeberry Finn* was their inability to match the handwriting of the men they pretended to be.[13] Once the source, usually American merchants in the Mediterranean, was verified, it was fairly easy for a reader familiar with the mercantile world to gauge its reliability. For example, the firm of John Bulkely and Son, of Lisbon, which had long been crucial to the Philadelphia-Mediterranean trade, was among the most prominent and reliable sources of information on Iberia. The Bulkelys frequently sent reports of news and warnings of potential danger to their many mercantile connections throughout the United States.[14]

Unfortunately, letters could also mislead. For example, the letter that McComb carried in 1785 was not exactly the earliest word that Americans had been captured by Algierians. Precedent must go to a letter sent from Alicante to a London merchant almost a month earlier reporting that "a large American ship from Boston . . . was captured by an Algerine cruiser . . . and the Captain and crew sent into slavery." This letter reached Boston at the same time as McComb, thereby reinforcing his news. Ironically, it was a false report written two days before the first American ship was actually captured. At about the same time, members of Congress fretted over a Virginia ship supposedly captured by Algerians, which later proved to have been taken and released by Moroccans instead.[15] Letters were only as accurate as the reports on which they were based. It is possible that these misleading reports came from confused seaman. At this time, a number of American crews just barely escaped capture by Algerians, sometimes abandoning their ships to the "pirates."[16] It is possible that the London merchant's informant heard of one of these incidents or saw a captured American ship and assumed that the crew had been sold into slavery. It is also possible that the London merchant was practicing a clever deception. Aware that letters on ship captures were widely read by the public, he might have placed this piece of false information in the hope that it would steer worried merchants away from hiring American ships and cause them to rely more heavily on supposedly safer English vessels. Such devious manipulation of the news would have come as no surprise to Thomas Jefferson, who in 1785 assured a New York correspondent that the Algerians had captured only two American vessels. "I mention this," he wrote, "because the English papers would make the world believe we have lost an infinite number."[17] The London merchant's letter, and others like it, were at the root of those devious newspaper reports.

In theory a nation's diplomatic service can act as a clearinghouse for information on foreign affairs. A well-run diplomatic corps can gather information more effectively than the mercantile network or even a private news service because of the breadth of its contacts and the power and prestige of the state. However, in the case of the United States the reality fell far short of this ideal, particularly in the 1780s. In 1785, at the time of the first captures, there were only two American officials permanently stationed in Mediterranean ports near Algiers—Robert Montgomery in Alicante and Richard Harrison in Cadiz. In addition, a merchant, Thomas Barclay, had been sent from Cadiz as a special agent to negotiate with the Emperor of Morocco. Three more distant American diplomats—William Carmichael in Madrid, Thomas Jefferson in Paris, and John Adams in London—would be involved in Algerian affairs because of their higher rank. Later, the Confederation Congress would send special agent John Lamb and his secretary, P. R. Randall to Algiers. Judging from his reports, Lamb, a former mule trader, was only semi-literate, and he proved to be as inept a diplomat as he was a writer.

Thus the early diplomatic network was not in a position to offer much that was new to the public or to the government in Philadelphia. In fact, newspapers may have put more stock in British diplomatic reports in these days. For example, Bostonians could read in their local paper that "some dispatches were received at the Secretary of State's office from Gibralter . . . [that] contain an account of the safe arrival there of upwards of twenty sail of ships from America, and two of which had been stopped and plundered by the Algerines."[18] The "Secretary of State" to which the item referred was at London rather than Philadelphia. One of the diplomats' major tasks was to pass on information from European newspapers to Philadelphia, and frequently they seem not to have known much more than they read in these journals. For example, Jefferson wrote to John Adams in 1785, "I observe in the Leyden Gazette . . . the extract of a letter dated Algiers April 15 which says that on the tenth of April an American vessel, the Clementina, Captain Palmer from Philadelphia, was carried in there by a cruiser." Jefferson had no way to verify the report, but he concluded that it was probably true because other details in the letter pertaining to American affairs were accurate. Similarly, Thomas Barclay passed on to Jefferson a report from a correspondent in Mogadore that Algerians planned to attack the coast of America. Barclay was skeptical but unsure. He wrote, "I cannot give entire credit to the account as the season of the year is far advanced, and we have not heard that any of the Algerine cruisers have passed the Streights [sic] of Gibraltar."[19] Clearly these officials were doing their best to gather intelligence, but they simply lacked the manpower to due an effective job. Members of Congress frequently complained that they learned more about Algiers from the newspapers than from these diplomats.[20]

Even in this early period, the diplomatic network could serve a useful purpose by confirming or denying other reports. For example, when wild rumors of Algerian ships in the West Indies were spreading through the United States, someone made public a portion of a report made by P. R. Randall, secretary to the American agent in Algiers, asserting that the Algerians sailed "very rarely to the Western islands, as many in America have imagined."[21] In addition, the differences between diplomatic correspondence and the merchant correspondence network or the sailors' grapevine should not be overemphasized. While diplomatic reports were "official," and therefore more trustworthy, diplomats, particularly at the consular level, were often merchants themselves. By the 1790s, America's most important diplomat in southern Europe, David Humphreys, was married to the daughter of one of the region's leading mercantile firms. Furthermore, diplomatic reports were carried in the same ships that carried sailors and merchant letters. All this information tended to become intermingled, thereby minimizing the uniqueness of diplomatic reports. Finally, diplomatic reports by necessity moved at the same speed as other reports and in fact often arrived later than "unofficial" information. To cite a somewhat later example, the first semi-official word of peace with Algiers, sent by the American consul at Malaga to a private citizen, became public in Boston in January of 1796, nine months after the first treaty rumors reached Massachusetts.[22] In this case the consular information was not new, but the prestige of the diplomatic network lent it an aura of certitude that earlier reports lacked.

The only relatively new technology of the period was the printed newspaper. Newspapers emerged in the English speaking world about a century before 1785, but they were not very prevalent in North America until the late eighteenth-century. While only seventeen newspapers operated in what is now the United States in the entire half century between 1690 and 1740, in the year 1785 alone twenty newspapers began operation in the new republic. By 1790 readers in the United States could choose from more than one hundred newspapers, and by 1800 that figure leaped to more than two hundred fifty.[23] Despite their novelty, newspapers did not represent a sharp break with earlier communications technology. Older forms—oral reports, merchants' letters, and the diplomatic network—frequently found their way into newspapers. The first report of the 1785 Algerian captures printed in American newspapers contained both a copy of the letter that McComb carried and a short summary of the information that he reported orally. Unlike their modern descendants, eighteenth-century newspapers were not staffed by reporters who collect, collate, and analyze information before presenting it in highly refined form to the reader. Instead, their closest modern counterpart might be moderated list serves, electronic bulletin boards in which information from all sorts of sources throughout the world is cut and pasted into a printed site by an individual who chooses what to print but does little to modify (or verify) the content.

When considered solely from the perspective of numbers printed, newspapers appear to be a puny portion of the early national public sphere. But two factors greatly augmented their importance. First, although it was rare for a publisher to print more than two thousand copies of any single newspaper, each copy typically passed through many hands, particularly in coffee shops which served practically as newspaper reading rooms. Furthermore, coffeehouse denizens could read these newspapers aloud to their less-literate comrades. Beyond this secondary and tertiary circulation, newspapers also served to reflect all aspects of the public sphere by setting into print letters, oral reports, and official documents circulating throughout the cities and across the Atlantic. The illiterate waterman might have heard firsthand the exact report that the merchant read in the day's newspaper while sitting in his coffeehouse. Neither of them needed to purchase or read the newspaper to be exposed to the information printed in its pages. In this way the newspaper served not only to record the day's news, but also to echo the vibrations of the public sphere.

Overall, then, the variety of late eighteenth-century media—oral reports, letters, official documents, and newspapers—allowed a wide swath of the population some exposure to the latest world news. Most importantly, in a day when print and oral culture merged together, literacy was not a barrier to receiving information. Not only were non-elites exposed to these media, but they also had an opportunity to influence them to some extent. The most notable group was the often-plebeian and frequently illiterate seamen described by one historian as "bearers of culture and information among far-flung groups and places."[24] Within urban centers, there can be little doubt that all sorts of people—male, female, rich, poor, black, white, literate, and illiterate—served as newsbearers, reciting the news that they read or heard to their friends and family every day. Of course, merely hearing or reciting news did not necessarily make these plebeians full participants in what Habermas termed the "public sphere" and contemporaries called the "Republic of Letters." It is generally understood that this term implies a *critical* discourse, that is reasoned analysis as well as repetition of tidbits of information.[25] In this sense many of the individuals who helped spread the news of the Algerian captures might be considered second-class citizens of the Republic of Letters. That is, they were aware of what went on in public discourse and could even contribute to the spread of news and information, but due to illiteracy, race, gender, or social status they may not have been able to participate fully in the critical analysis which was primarily the domain of the literate, white, male, bourgeoisie. But this level of analysis could not occur without access to basic facts; if a ship was captured in the Mediterranean and no one heard (or read) about it, it could have no effect. Furthermore, the content of the earliest news reports inevitably influenced subsequent critical analysis. Thus, by transmitting basic information these second-class citizens played an important role in shaping the Republic of Letters.

Bringing the State in, 1793-1794

Reports of the second group of captures, in 1793, reached the United States much more quickly than did those of the earlier round. In 1785 McComb's letter first arrived in the United States roughly three months after the captures of the *Maria* and the *Dauphin*. It took another month for this news to work its way down the coast to Charleston, meaning that most Americans learned of it three to four months after the actual events. By contrast, word of the 1794 captures arrived just shy of two months after the events occurred, making its way to Charleston in less than two weeks so that the vast majority of the country was aware of the captures in just a little more than two months after they occurred. This improvement was due, in part, to a coincidence. In 1793 a Swedish ship saw the captures as they occurred off the Iberian coast, while in 1785 first word came from Algiers, meaning not only that the news had farther to travel but that it got a slower start. The coincidence, however, was only part of the difference. Between 1785 and 1793 the United States had ratified its current Constitution and undertaken construction of a more powerful central government. This more powerful state played an important role in facilitating better communication in the late 1790s.

Between 1785 and 1793 the United States began to develop a diplomatic corps. It was still small and not particularly professional in 1793, but it is clear that it made a real difference in expediting the transfer of information. At almost every step in the process of moving the news of the captures from Gibraltar to the United States, American diplomats played a vital role. Their participation gave the news an official imprimatur, making it far more believable to the American public than McComb's letter, the veracity of which had been doubted by many at the time. The key player was David Humphreys, Minister to Portugal, who happened to be in Gibralter on October 6, 1793. On that day a dispatch from the English consul to Algiers arrived in Gibralter with news of a twelve-month truce between Algiers and Portugal.[26] The news was troubling because Americans and others depended on Portuguese hostility to the Algerians to act as a check, keeping the "pirates" bottled up inside the Mediterranean. Without the Portuguese as a counterbalance, Algerians could enter the Atlantic with impunity and capture scores of unsuspecting vessels. Thus it was imperative that American ships be warned of the truce before they were taken unaware by Algerian cruisers.

Humphreys undertook heroic measures to spread the word. He immediately sent out warning letters "to all governors, magistrates, officers civil, military, and others concerned in the United States of America."[27] The recipients of these warnings in turn alerted American ship captains to the danger.[28] At the same time, Humphreys hired a small Spanish ship to cruise the Streights of Gibralter for fifteen days to inform any passing American vessels of the truce.[29] Humphreys also made repeated efforts to get the news of the truce home to America as quickly as possible. He personally sent out at least five separate copies of his letter to Jefferson in Philadelphia in the hope that one might get through. In addition, he sent word to the American consuls in Lisbon and Cadiz to charter neutral vessels in an effort to move the news even more rapidly and safely.[30]

Humphreys' dispatches mingled with even more disturbing news. On or about October 12, Algerian cruisers stopped and searched a Swedish vessel sailing off Portugal's southwestern extremity. Luckily the Algerians respected Sweden's neutrality and allowed the ship to proceed. But others were less fortunate. The Swedes noticed that the Algerians had four American vessels and one Genoese ship in tow and they appeared to be in the process of capturing two more American vessels. The Swedes rushed back to Lisbon, bringing their news into that port by October 15, just as the American consul, Edward Church, was attempting to comply with Humphreys' order to charter an express to America.[31] Church and the American merchants in Lisbon were near panic, afraid that many more of their countrymen would soon be captured and "enslaved." Meanwhile, the American consuls continued making arrangements to get the news to the United States. In Lisbon, Church and the New York merchant Schuyler Livingston negotiated to charter a Swedish ship, the *Mary*, which they hoped could rely on Swedish neutrality to cross the Atlantic undisturbed.[32] In Cadiz, Consul Peter Walsh, unable to secure a neutral vessel, negotiated with the captain of a New York ship, the *Two Sisters*. In the midst of these negotiations word of the American captures arrived in Cadiz, allowing Walsh to supplement the truce warning with this new information. Now, a little more than a week after Humphreys sent out his first warning, all three ships were ready to sail for America. In addition, about one week later, Church chartered a second neutral carrier, the Danish *Statdt Altona*, to carry duplicates of his dispatches.[33]

The *Mary* reached her destination first. Livingston landed in New York and began to spread the warning on December 9 just two months and a day after Humphreys' dispatch had first arrived in Lisbon. The news spread rapidly throughout America, arriving in Philadelphia by December 11, Annapolis by December 19, and Charleston by December 21. In Boston, it was preceded by a letter sent by Lisbon merchants on their own initiative, which arrived on December 16. Rather anticlimactically, the *Twin Sisters* arrived in New York on December 17 and the *Stadt Altona* on December 23, both carrying what had become yesterday's (or last week's) news.[34]

The diplomatic corps was not the only national institution that expedited information diffusion. The development of a national postal service intent on encouraging rapid and inexpensive distribution of newspapers no doubt made a difference once the news crossed the Atlantic.[35] The historian Richard John has shown in impressive detail the impact of the Post Office Act of 1792 on what he calls the American "informational environment." By admitting newspapers into the mails at low rates, this act did much to spread the news in early national America. John finds that the number of post offices per capita doubled in the decade between 1790 and 1800. Given that newspapers made up approximately 95 percent of the weight of the mail, the increase in delivery service led directly to an increase in access to the news. Furthermore, the government allowed newspaper editors to receive free copies of all of their counterparts' publications through the mails. Because the great bulk of newspaper articles were cribbed from other newspapers, this privilege was instrumental in facilitating rapid dissemination of reports such as those from Algiers.[36] In short, the emergence of a centralized national government was a key factor in expanding the geography of knowledge.

It was not just the apparatus of the nation state that fostered the diffusion of information. The very achievement of independence conditioned Americans to be more inquisitive and concerned about foreign affairs. Before the Revolution they could be relatively complacent, assured that the mother country would take care of most problems as only a great world power can. The conflict with Algiers, as the new nation's first major diplomatic crisis, revealed the post-colonial United States as an inexperienced and almost pitifully weak player on the world stage. At least one report from the mother country chortled at the new nation's hard lesson. "How different is the case now from what it was," observed one London newspaper, recalling that before independence "the American vessels could sail secure, and had no other enemy to dread but the wind and waves!"[37] This new situation necessitated a heightened awareness of foreign affairs on the part of Americans. The necessity was augmented by the fact that the new nation had set itself up as a democratic republic which required an informed citizenry.[38] If any measures against Algiers were to succeed, public support would ultimately be necessary. Public isolationism could be fatal, as John Adams noted when he complained to Jefferson that "a disposition seems rather to prevail among our citizens to give up all ideas of navigation and naval power, and lay themselves consequently at the mercy of foreigners, even for the price of their produce." Far from apathetic, the public ultimately proved concerned enough about events in the Mediterranean that a solid majority of Representatives in Congress felt justified in supporting the establishment of a navy in 1794 on the grounds that it would enable the United States to neutralize the Algerians.[39] In the case of the United States, at least, regardless of technological progress the emergence of the liberal state and its need for an informed citizenry broadened the geography of knowledge. Quite likely a similar process occurred in revolutionary France as the fate of the new republic became enmeshed with world affairs. Liberal republics require a great deal more information than states in which decision making is limited to a small cadre.

Living in a Wider World

The American reaction to the Algerian crisis helps to illuminate some of the implications of the broadening geography of knowledge. Perhaps most clearly, it reveals what it meant to live in the expanding world of trade that emerged during this period of merchant capitalism. Expanding capitalism was both a cause and effect of information diffusion. As Habermas himself noted, the traffic in news developed in connection with the needs of commerce.[40] That is, increased availability of information about prices and conditions in distant ports created more efficient and therefore more profitable markets. In turn, broader markets prompted the need for more information, creating a secondary market for world news that was filled by the emerging newspaper press. The Algerian case suggests that these developments created new risks and anxieties for the individuals who experienced them. As if hazards of daily life such as disease, malnutrition, fire, and ongoing warfare with neighboring Indians were not worrisome enough, Americans now had to be concerned with the activities of pirates living far across the Atlantic. Jefferson worried about his daughters' safety. Merchants worried about their ships and cargoes. Sailors' wives worried about their husbands. Some Americans worried about the prospect of an Algerian invasion of the United States.[41] American officials worried that the North African pirates would shut down the new nation's promising Mediterranean commerce. Then, as now, globalization created risks as well as benefits.

The increased diffusion of knowledge also influenced individual and group identity formation. The cosmopolitanism associated with the Enlightenment played an important role. Statesmen like Franklin and Jefferson bridged the Atlantic, finding themselves nearly as comfortable in Paris as in Philadelphia. The Algerian crisis prompted a cosmopolitan like Jefferson to consider a multinational solution: an alliance with a number of small European states, including Naples, Portugal, Tuscany, and perhaps later, Russia. Cosmopolitanism was also evident at lower social levels. Blinckhorn, for example, was a sort of citizen of the world, aware of world affairs and able to profit from them in the United States. Another grifter, one James Joshua Reynolds, regaled a group of gentlemen in Greenock, Scotland, about his tribulations as a slave in Algiers, where he claimed to have met many other lost sailors including one by the name of Spence. Spence's parents read the account in a Virginia newspaper and briefly gained hope their son was still alive. However despicable Reynolds' motivations may have been, his knowledge of world affairs, right down to the names of missing sailors, is impressive. It reveals the widespread diffusion of information among sailors throughout the Atlantic and Mediterranean. A man like Reynolds, with his trove of interesting (albeit often false) information could be at home practically anywhere. Such cosmopolitanism has prompted one historian to suggest that eighteenth-century seaman was not only a "man of the world," but part of an international proletariat.[42]

Yet, for all its cosmopolitanism, this was also an age of burgeoning nationalism. One expert notes that "to a surprising extent American seamen remained Americans during the Revolution," motivated in part by hatred of the English.[43] In fact, far from being antithetical forces the cosmopolitanism inherent in the expanding geography of knowledge reenforced the incipient nationalism of this era for the general public as well as for sailors. This sense of nationalism should not be confused with jingoism; for the most part it was defensive and fearful. The Algiers crisis revealed to a shocking extent America's post-colonial frailties. After all, the new nation now appeared too weak and too diplomatically inept to deal with what was barely a third-rate power. The Algerian crisis made it very clear to Americans that they were no longer protected by Britain; if they were it would never have occurred in the first place. This was not a very reassuring lesson, but anxiety-provoking as it may have been it made the perilous reality of nationhood much clearer.

The Algerian crisis also prompted the first outpouring of angry nationalistic righteousness. It was directed not against Algiers as one might expect, but against England. Americans believed they saw English machinations behind the Algerians' actions, most likely an effort to get even with the former colonies. As early as 1785 Ralph Izard wrote Jefferson from South Carolina that "it is said that Great Britain has encouraged the piratical states to attack our vessels." Several delegates to Congress suspected the same thing. By the 1790s this supposition was gospel for many Americans who believed Britain had engineered the 1793 truce between Algiers and Portugal in order to make it easier for Algerians to capture American ships. At a well-attended 1794 mass meeting, New York's Republicans condemned "the embarrassments and injuries offered to the commerce of our country by Britain and her savage allies," a reference to Native Americans as well as Algerians.[44] Whether or not such allegations were true (and subsequent revelations cast doubt on their veracity) they did serve to create a sense of a common destiny in a still rather inchoate nation only a few years removed from colonial status. In this way the Algerian crisis led individuals to identify more closely with their nation at the same time that it represented increased globalization.

For the historian this is a useful tale for understanding the connections between enlightenment, nationalism, world trade, and the expanding public sphere. There can be no denying that, beyond this historical context, it also has a very modern ring. No doubt many Americans today can identify with the experience of their distant ancestors when confronted with frightening violence originating in poorly understood, distant places. Without a doubt, many Americans have had their geography of knowledge expand since September 2001, as once unheard place names such as Kabul, Peshawar, and Kandahar became all too familiar in their daily lives. Viewed from their perspective, this phenomenon must seem less a part of an unfolding centuries-long process of globalization than a new, and alarming development. For these Americans, as for their predecessors who experienced the Algerian crisis, such events bring new understanding of the individual's place in a larger geography that can be revealed through its hazards as well as its opportunities.

Notes

1. David Jaffe, "The Village Enlightenment in New England, 1760-1820," *William and Mary Quarterly, 3rd Series* 47 (1990), 327-346.

2. Marcus Rediker, *Between the Devil and the Deep Blue Sea* (New York: Cambridge University Press, 1987); Brooke Hindle, *The Pursuit of Science in Revolutionary America, 1735-1789* (New York: W. W. Norton, 1974); Bruce Sinclair, *Philadelphia's Philosopher Mechanics: A History of the Franklin Institute, 1824-1865* (Baltimore: Johns Hopkins University Press, 1974); Edwin Wolf II, *The Book Culture of a Colonial American City: Philadelphia Books, Bookmen, and Booksellers* (New York: Oxford University Press, 1988); Cathy Davidson, *Revolution and the Word: The Rise of the Novel in America* (New York: Oxford University Press, 1986); Rosalind Remer, *Printers and Men of Capital: Philadelphia Book Publishers in the New Republic* (Philadelphia: University of Pennsylvania Press, 1996); Albrecht Koschnik, "The Democratic Societies of Philadelphia and the Limits of the American Public Sphere, Circa 1793-1795," *William and Mary Quarterly, 3rd Series* 58 (July, 2001), 615-636; John L. Brooke, "Ancient Lodges and Self-Created Societies: Voluntary Association and the Public Sphere in the Early Republic," in Ronald Hoffman and Peter J. Albert, eds., *Launching the "Extended Republic": The Federalist Era* (Charlottesville: University Press of Virginia, 1996), 273-377.

3. Jurgen Habermas, *The Structural Transformation of the Public Sphere: An Inquiry into a Category of Bourgeois Society* (Cambridge, Mass.: MIT Press, 1989); Craig Calhoun, ed., *Habermas and the Public Sphere* (Cambridge, Mass.: MIT Press, 1992); Michael Warner, *The Letters of the Republic: Publication and the Public Sphere in Eighteenth-Century America* (Cambridge, Mass.: Harvard University Press, 1990); David Waldstreicher, *In the Midst of Perpetual Fetes: The Making of American Nationalism, 1776-1829* (Chapel Hill: University of North Carolina, 1997); David Shields, *Civil Tongues and Polite Letters in British America* (Chapel Hill: University of North Carolina, 1997); Lawrence A. Peskin, "From Protection to Encouragement: Manufacturing and Mercantilism in New York City's Public Sphere, 1783-1795," *Journal of the Early Republic* 18 (Winter, 1998), 589-615.

4. *Boston Gazette,* Dec. 5, 1785, 1; Mathew Carey, *A Short History of Algiers* (Philadelphia: Carey, 1794).

5. Charles E. Clark, *The Public Prints: The Newspaper in Anglo-American Culture, 1665-1740* (New York: Oxford University Press, 1994), 215-242; Frank Luther Mott, *American Journalism: A History of Newspapers in the United States Through 250 Years, 1690-1940* (New York: The Macmillan Company, 1941), 49-53.

6. Carl E. Prince, et al., eds., *The Papers of William Livingston* (New Brunswick, N. J.: Rutgers, 1979) V: 410-412, 416-417.

7. *Salem Gazette,* Oct. 25, 1785, 3.

8. *Pennsylvania Gazette*, June 11, 1794 (Item #79704); (Charleston, S.C.), *Columbian Herald* August 30, 1787, 2.

9. Charleston, (S.C.), *Evening Gazette* April 7, 1786, 3. For later but similar report see also *Columbian Herald*, Jan. 22, 1794, 2 and Annapolis, *Maryland Gazette* Feb. 6, 1794,

2.

10. Julian P. Boyd, ed. *The Papers of Thomas Jefferson* (Princeton, N.J.: Princeton University Press, 1953) X: 653-654. Jefferson also looked into the case of a Captain Greene whose father, for unknown reasons, believed he might be among the captives in Algiers.

11. Richard Brown, *Knowledge is Power: The Diffusion of Information in Early America, 1700-1865* (New York: Oxford University Press, 1989), 114-116.

12. For examples of oral reports of letter contents see *Boston Gazette* Nov. 23, 1789, 3 ("the emperor of Morocco lately sent a vessel to Madeira, with information to Mr. Clark"); *Boston Gazette*, June 27, 1793 ("Captain Hoover saw a letter from the American consul at Malaga."). Letters passing on news received orally were ubiquitous.

13. Samuel Langhorne Clemens, *Adventures of Huckleberry Finn* (New York: W. W. Norton, 1961), 158-159.

14. Reprinted in *Maryland Gazette*, June 19, 1794, 3; *Columbian Herald*, April 16, 1794, 3. For other Bulkely letters see *Boston Gazette*, Oct. 17, 1796, 3; *Maryland Gazette*, January 30, 1794, 2. On the Bulkelys' Philadelphia connections see Thomas M. Doerflinger, *A Vigorous Spirit of Enterprise: Merchants and Economic Development in Revolutionary Philadelphia* (Chapel Hill, University of North Carolina Press, 1986), 62.

15. *Boston Gazete*, Oct. 31, 1785, 4; Paul H. Smith, ed., *Letters of Delegates to Congress* (Washington, D.C.: Library of Congress, 1994) 22: 230, 686-687.

16. For example, see *Columbian Herald*, Nov. 21, 1785, 2, and *Maryland Gazette*, Nov. 17, 1785, 2.

17. *Papers of Thomas Jefferson* 8: 559.

18. *Boston Gazette*, Nov. 16, 1789, 2.

19. *Papers of Thomas Jefferson* IX: 645-646 and X: 536-536. On the operations of the early State Department, see Leonard White, *The Federalists: A Study in Administrative History, 1789-1801* (New York: The Free Press, 1948), 128-144.

20. *Letters of Delegates to Congress* 23: 14, 21, 32, 37.

21. *Boston Gazette* Sept. 25, 1786, 2. The newspaper report does not identify Randall, but the wording is identical to his report dated May 14, 1786, in *Jefferson Papers* IX: 535.

22. *Boston Gazette*, Jan. 4, 1796, 3; *Maryland Gazette*, Mar. 24, 1796, 1. For an early, and inaccurate rumor of the completion of the treaty, see *Boston Gazette*, March 16, 1795, 2. For an account of the negotiations see Irwin, 69-81. For a warning to the public from Secretary of State Timothy Pickering that the treaty was not yet in operation see *Pennsylvania Gazette*, June 15, 1796 (Item #81203).

23. Charles E. Clark, *The Public Prints: The Newspaper in Anglo-American Culture, 1665-1740* (New York: Oxford University Press, 1994) 267-268; Jeffrey L. Paisley, *"The Tyranny of Printers": Newspaper Politics in the Early American Republic* (Charlottesville: University Press of Virginia, 2001), 403-404.

24. Rediker, *Between the Devil and the Deep Blue Sea*, 203.

25. See Warner, *The Letters of the Republic,* 30-33, for a particularly cogent explanation.

26. *Papers of Thomas Jefferson*, 27: 196-197; James Simpson to Jefferson, Oct. 8, 1793, State Department Consular Dispatches (hereafter SDCD) Gibralter Series (paraphrased in Papers of Thomas Jefferson, 27: 224).

27. David Humphreys, "To all governors . . ." Oct. 8, 1793, SDCD Cadiz Series; *Papers of Thomas Jefferson* 27: 196-197; Humphreys to Edward Church Oct. 8, 1793, SDCD Lisbon Series; Michael Murphy to Sec. State, Oct. 11, 1793, SDCD Malaga Series.

28. Michael Murphy to Sec. State, Oct. 11, 1793, SDCD Malaga Series; Samuel Calder to Mr. David Pearce, Jr. Dec. 4, 1793, SDCD Algiers Series.

29. Humphreys to Sec. State, Oct. 10, 1793, and Oct. 13, 1793, Dispatches of U. S. Ministers to Portugal.

30. Humphreys to Edward Church, Oct. 8, 1793, Dispatches of U.S. Ministers to Portugal; *Papers of Thomas Jefferson*, 27: 222-223.

31. *Pennsylvania Gazette*, Dec. 11, 1793 (Item #79299); *Maryland Gazette*, Dec. 19, 1793, 2.

32. *Papers of Thomas Jefferson* 27: 230-233.

33. Peter Walsh to Thomas Jefferson, Oct. 16, 1793 and Oct. 17, 1793 (2 letters), all in SDCD Cadiz Series.

34. *Pennsylvania Gazette*, Dec. 11, 1793 (Item #79299), and Dec. 24 1793 (Item #79353); *Boston Gazette* Dec. 16, 1793, 3; *Maryland Gazette*, Dec. 19, 1793, 3.

35. Richard R. John, *Spreading the News: The American Postal System from Franklin to Morse* (Cambridge, Mass.: Harvard University Press, 1995), 25-63.

36. Richard John, "Recasting the Information Infrastructure for the Industrial Age" in Alfred D. Chandler Jr. and James W. Cortada, *A Nation Transformed by Information: How Information has Shaped the United States from Colonial Times to the Present* (New York: Oxford University Press, 2000), 59-61.

37. *Boston Gazette*, May 1, 1786, 3; *Columbian Herald*, March 5, 1787, 2; *Evening Gazette*, Dec. 15, 1785, 2.

38. On this concept, see Richard D. Brown, *The Strength of a People: The Idea of an Informed Citizenry in America, 1650-1870* (Chapel Hill: University of North Carolina Press, 1996).

39. *Journal of the House of Representatives of the United States* (Vol. 2, 1794), 86-87.

40. Habermas, *The Structural Transformation of the Public Sphere,* 21.

41. Robert J. Allison, *The Crescent Obscured: The United States and the Muslim World, 1776-1815* (New York: Oxford University Press, 1995) 3-7; *Boston Gazette*, Dec. 12, 1785, 3; May 1, 1786, 3; *Pennsylvania Gazette,* April 5, 1786 (Item #72668); *Evening Gazette*, Mar. 6, 1786, 2.

42. Reddiker, *Between the Devil and the Deep Blue Sea*, 10-76, 288-298.

43. Jesse Lemisch, "Jack Tar in the Streets: Merchant Seamen in the Politics of Revolutionary America," in *In Search of Early America: The William and Mary Quarterly 1943-1993* (Richmond, Virginia: William Byrd Press, 1993), 131.

44. *Papers of Thomas Jefferson* VIII: 195-196; *Letters of the Delegates to Congress*, 22: 683, 692; 23: 456; New York *Diary*, Feb. 25, 1794, 2, and Feb. 27, 1794, 2.

Chapter 9

Finding the There There: Local Space, Global Ritual, and Early Cold War Berlin

Paul Steege

Introduction

The start of the twenty-first century finds historians wrestling with the collapse of the discipline's "grand narratives" and the uneasy ambiguities of increasingly plural histories in which multiple perspectives offer unsettled and multi-vocal takes on the past. While some have critiqued any move from the macro (grand narratives and large structures) to the micro (the history of everyday life) as a retreat into the anecdotal,[1] the real power of a microhistorical approach depends not on shift of focus to a particular, narrowly circumscribed, material location, but rather on an alternative framing of an historical reality that depends on the tension between macroscopic and microscopic levels of symbolic and material meaning. This tension can only be apprehended in a liminal space—"betwixt and between"—that enframes the nexus produced by the encounter between these contested meanings. While this approach endeavors to challenge simplistic notions of top-down imposition of power and meaning, it offers more than just a history "down below." The struggle to *locate* this history, to take seriously both individual agency and the complementary and contradictory structures of meaning that explain it, ultimately reasserts the human scale of events that also carry world implications.

Historical writing on the Cold War has tended to focus on questions of temporality and causation: When did it start? How long did it last? Who was to blame? The geography of the Cold War, however, has evoked significantly less scholarly concern, the "fact" of Cold War "battlegrounds" simply providing a list that served as a neat shorthand for the defining *events*: Korea and Vietnam (the respective wars), Cuba (the missile crisis), or perhaps most iconic of all, Berlin (the Blockade and later the prolonged face-off across the Wall).[2] These battlegrounds marked the symbolic geography of the Cold War, enframing the productive tension of material and symbolic conflict in multifaceted locations. But this process of delineating these locations—an act of meaning production central to the Cold War itself—remains largely unconsidered.

Typically, the word battleground implies a location, a material space in which or upon which the combatants clash. While the terrain in this space may impact the battle (in physical terms, the individual may stumble, the army unit be stuck in a ditch), it most often seems to function as little more than the setting or even backdrop for the *real* action, the battle itself. Indeed, to take it a step further, this location seems not so important in and of itself but only as it is fought over or as it becomes the site of a battle, whose significance, in turn, rests on factors and issues most likely independent of this particular location. At first glance, Cold War Berlin effectively serves to illustrate this perspective, perhaps nowhere better than John F. Kennedy's June 1963 speech in front of West Berlin's city hall. Speaking to a packed square, he called on the world to "come to Berlin" if they had any questions about the nature of the Cold War. He ended his remarks, "All free men, wherever they may live, are citizens of Berlin, and, therefore, as a free man, I take pride in the words, '*ich bin ein Berliner* [I am a Berliner].'"[3] The massive crowd responded with thunderous applause to the president's celebration of the city and its heroic place in a global struggle. But, if "all free men" could claim the status of Berliners, then what of those who really lived there? Even more significantly, what of the actual, material city, Berlin? What connection did it have to the iconic vision with which Kennedy so powerfully summed up the Cold War?

This question, suggested by the rhetoric of an American president, in fact raises much broader issues. First, it asserts the need to take seriously the messy, ambiguous relationship between location, politics, and the symbolic constructions and representations of power that helped constitute the Cold War as a world conflict. More fundamentally, it challenges pat formulations of the links between local action and global explanatory frameworks and demands that we reconsider simplistic notions of historical agency, in which a "battleground" is simply a *site* for great power contestation. It is not enough, for example, to assume that the Cold War was a Soviet-American conflict imposed upon the world. It remains necessary to consider the particular contexts in which locations assumed specific significance for this clash and the way it came to be defined.

Locating Symbolic Meaning

At the root of this effort to craft precise formulations of the struggles in and for particular locations rests the need to delineate a clear sense of the relationship between the exercise of power and the symbolic meanings with which those acts come to be defined and/or experienced. On some level at least, symbols function as structures of domination.[4] The extent to which they operate successfully determines whether or not the structures they define are perceived as legitimate; but the structures of domination remain regardless. According to Pierre Bourdieu, even in the case of willing and voluntary submission to the force of (legitimate) authority, it is possible describe such domination as "symbolic violence." Although authority demands that its subjects submit, they do not perceive their actions in that way.[5] Indeed, Bourdieu argues that as submission is necessarily a "cognitive act," even the most brutal "relationship of power" is also a symbolic relationship, that is, a relationship that depends upon more than material coercion.[6] The persons subject to the exercise of power are necessarily participants in their own subjugation to authority and the symbolic paradigms under which that authority operates. The key to this process is a certain distancing from the harsh realities of submission, a masking of the relationships of power.[7] The interaction or exchange, which categorizes this exercise of power, depends on a cooperative denial of the process itself. One acts *as if* the action does not mean what it does.[8] This action in the subjunctive mood anchors the whole relationship between the material and the symbolic. While the symbol to some extent conceals the (material) reality of the action, it does not deny the agency of the individual "on the ground."[9] The possibility of independent human action depends on the stance that the individual assumes—an interactive process imbedded explicitly in a material and symbolic geography.[10]

For the anthropologist, Victor Turner, the subjunctive mood, Bourdieu's *as if*, marks a particular location "betwixt and between" that serves to define both the individual and the place of that individual within or with respect to human society. Turner first described this liminal state in relationship to rites of passage among the Ndembu people of south-central Africa. He noted that participants in these rites, especially initiates into adulthood, found themselves in this in-between condition that is at the same time apart from and distinct for their society.[11] Yet this in-between-ness is not only being part but not part of society (after all, only by means of the rite does the initiate fully enter adult society) but also constitutes an alternative form of social organization that Turner calls "antistructure." The ritual processes exist outside of ordinary time and rest on principles different from those of everyday social life.[12] On the one hand this promotes an intense sense of community among the initiates, but it also permits them to depart from the constraints of normal social structure and engage forces that transcend the moral limitations of the regular social order. Even more explicitly:

> Liminality, seen from this angle, is pure potency, where anything can happen, as everything is stated to have happened to the primordial powers; where immoderacy is normal, even normative; and where the elements, the basic building blocks of culture . . . are released from their customary configurations and recombined in often bizarre and terrifying imagery.[13]

This power is limited by participants knowledge "that this is a unique situation," but the possibilities inherent to it are not entirely hazard-free.[14] The meanings and explanations that the initiates encounter or experience in this liminal context can provoke unsettling challenges to the normal structure, as well. Ostensibly, the hazards of "boundlessness" are constrained by the broader structures of the society to which this rite is designed to contribute, but such ideas can, taken seriously, be potentially revolutionary. For ultimately, the power of authority and of society more generally depends on it being taken seriously for what it claims to be and/or represent.[15] One could perhaps say the same thing about the Cold War as well.

Circumscribing the Cold War

The years since the fall of the Soviet Union have produced a number of declarations celebrating the final end of the Cold War but also, in some policy-making circles, a kind of nostalgia for the presumed clarity of that conflict.[16] In an era of "rogue states" and global terrorist networks, that earlier time in which international conflict seemingly operated under the domineering influence of two opposing superpowers perhaps looks more readily comprehensible and perhaps even more stable.[17] Yet, as the historian Guy Oakes has noted, "[t] he declaration of the end of the Cold War became an instant media cliché, whose banality obscured the institutionalization of Cold War culture in American life—a process that cannot be expected to end with the dramatic finality of the fall of the Berlin Wall or the collapse of the Soviet Union."[18] His point raises the possibility that the structures and conceptualizations that helped characterize the Cold War are not as past as many might believe but also suggests that an analysis of this particular case offers an explanatory framework that explodes any specific temporal or geographic location. Ultimately, such an exploration must necessarily focus on the construction of meanings. Rather than concentrating solely on causation, a more thorough assessment of these meanings depends on the elaboration of a functional history of the Cold War.

At its core, what drove the Cold War: thousands of nuclear weapons, a military industrial complex, or some sort of deeply imbedded Cold War culture? Even as Oakes describes the institutionalization of this culture in his analysis of the American government's 1950s decision to promote a vision of civil defense, he implies that it cloaked something else (a policy of nuclear deterrence). Although, the "imaginary war" to which the title of Oakes's book refers was the fictionalized vision of World War III pitched for the American public, he nonetheless draws a decisive line between the façade and the reality, between the Cold War for public consumption and the Cold War, which those in power discussed in private. Thus, even this particular cultural perspective suggests that ultimately the Cold War remained a superpower conflict that they imposed upon various publics around the world, a struggle defined from the top down.[19] This presumption that what was going on during the Cold War could be so readily explained (if perhaps not so easily countered) is quite simply misplaced. To a large extent, the confidence in the explanations hinges on the oversimplicity of questions asked of Cold War policies, ideas, and symbolism. There remains too great of a disconnection between Cold War symbolism and the material realities of Cold War action. From this perspective, the symbols explained, justified, or perhaps even incited the action, but they operated on a different plane, a plane of discourse or ritual that somehow remained separate from the *real* meaning of the Cold War. Thus, in trying to explain the Cold War, scholars sought to uncover more fundamental motivations, the drive for political power (Soviet expansionism) or economic domination (American capitalist imperialism), to explain what lay at the root of the crisis. The symbols that provided the public face of the Cold War presumably cloaked this underlying motivation, one that was by extension also more real.[20] However, the formulation of this "disconnect" between Cold War meaning and Cold War causation misses the point. In contrast, I argue that by turning more explicitly to the locations in which this tension (and even contradiction) manifests itself, the functioning of this symbolic geography of meaning production will also become explicit.

The Berlin Blockade was the first overt clash of the Cold War, and as such it played a dramatic part in shaping its iconography. According to the standard accounts, in an effort to halt the formation of a separate West German state or at least to try to force the three western powers from Berlin, the Soviet Union in June 1948 imposed a total blockade on all land transportation into the city's western sectors. For almost a year, an Anglo-American airlift ferried food and coal into the city, sustaining the West Berlin population and eventually forcing the Soviets to capitulate. After a series of secret negotiating sessions, Soviet and American representatives produced an agreement to end the blockade in May 1949, an agreement that helped stop the conflict short of a war that neither side wanted but also effectively paved the way for the establishment of two separate German states later that fall. Here, it would seem, one can easily apprehend the basic and readily decipherable reality of Cold War struggle: a Soviet power-play that overreached, a western tactical counter-blow (backed, of course, by the unspoken but ever-present nuclear monopoly), and the city of Berlin, the symbolic prize in the geo-strategic clash.

But if we cast another look at the Blockade's resolution, it is possible to see this explanation as insufficient. The problem with this account is that as far as Berlin goes, "there's no there, there."[21] The choices that officials made to address the situation in Berlin failed to account for the complex local situations, of which they generally had little or no awareness and which operated in ways that had little connection to the policies' intended outcomes. For fifty years, scholars and the general public alike described a nearly year-long Soviet blockade of West Berlin as a ruthless assault on the West's Achilles heel that totally cut off the western half of the city and threatened its population with starvation. Much recent scholarship has recognized that the Soviet blockade was at best incomplete and that the Anglo-American airlift was never the sole supply line for Berlin's western sectors. Instead, in the preceding years, ordinary Berliners had already developed trade and supply networks that transcended sector and zonal boundaries and facilitated the city's material survival.[22] The airlift did not achieve the western victory in the Blockade, at least not by doing what it claimed to be doing—supplying the material needs of Berliners. The decisive material battle was fought in the black markets and on the trains that Berliners rode to forage for food and fuel in the surrounding Soviet zone. This material battle centered on the difficult struggle to survive in a destroyed city, a struggle in which Berliners were the principal agents.[23] While the events in Berlin did not occur in a vacuum, the formulation of the city's symbolic resonance for an international Cold War likewise could not exclude the content of that Berlin location. Those events can make sense (prove meaningful) only in connection to (and in tension with) the particular conditions in the historical location of late 1940s Berlin. Indeed, these symbolic formulations are central to the functioning of the Cold War, but their seeming detachment from the material location of the conflict is illusory. The danger remains—for us as well as for the policy-makers whose decisions allegedly shaped the events at the time—that the density of the interaction between the material and the symbolic gets lost.[24] In this particular case, Berlin comes to exist only as a cipher, an essentially empty space in which, by chance as it were, a Cold War battle took place.

In purely physical terms, Berlin entered the postwar world as a city of rubble in which its inhabitants struggled to find food and fuel to cope with the harsh demands of that postwar urban existence. Surviving remained Berliners' principal activity, and this "reality" must underlie all other social, political, and symbolic analyses. In response both to this material crisis and to competing desires to assert and exercise power, the various great powers that occupied Berlin acted *as if* they were in a position actually to control the day-to-day events on the ground in the city. This *as if* helped to define the city as a multifaceted, liminal space (part but not part of the surrounding Soviet Zone, part but not part of the West, behind but not behind the Iron Curtain) in which a "temporary" contest within the city helped to craft a unique collaboration between the great powers, especially the Soviet Union and the United States, that helped sustain and justify their positions of abnormal (symbolic) power in a contest (the Cold War) that remained itself oddly liminal, in a way outside of both the course of time and detached from particular material and historical locations. A return to the events of late spring and early summer 1949 will help to muddle our sense of what was happening in and around Berlin at the time of the end of the blockade. By reviving a sense of the messiness "on the ground," I hope to clarify the functioning of this increasingly ritual conflict that depended upon the tension between macroscopic and microscopic enframings of meaning. The elaboration of this particular complex of contested meanings resonates far beyond any particular Cold War past, suggesting that the effort to unpack the historical meanings of international political conflict must be imbedded in the symbolic geography of a particular location.

Macroscopic Visions of Cold War Berlin

Most accounts of the Blockade have emphasized the increasing economic burdens, which its restrictions placed on the western sectors: plant shutdowns, layoffs, and growing unemployment.[25] These effects reflected the perception of Berlin as object of a total Soviet blockade. While such occurrences were a fact of life for many Berliners, the overall Berlin economy continued to function remarkably well.[26] Although the victorious allies divided postwar Germany into four separate occupation zones—Soviet, American, British, and French—they initially planned to treat the former German Reich as a single economic whole. Berlin, located over one hundred miles inside the Soviet Zone was technically not part of any zone and was itself divided into four occupation sectors. When optimistic visions of postwar political cooperation began to break down, these multiple boundaries suggested multiple obstacles to any efforts Berliners might undertake to rebuild their city and its economy. Nonetheless, Berlin remained one city in spite of its administrative divisions. Personal, economic, and political links did not stop at sector boundaries, even after the blockade had supposedly hardened them considerably. Most fundamentally, Germans in all zones shared a common encounter with real and sustained material scarcity. Nearly four years after the end of the World War II, almost all Germans still struggled to find enough to eat. This generally meant "organizing" foodstuffs above and beyond the "official" ration, which, even if wholly available, did not meet normal nutritional requirements. In East and West, supplemental food sources drove individuals' decisions about time management, work attendance, and organizational membership—all decisions that crossed both geographic and symbolic boundaries. It often made more sense to forego a day's wages to forage in the countryside (explicitly a different kind of space—the Soviet Zone) or trade on the black market.[27] The Berlin schools even included scrounging (*Hamstern*) in a list of reasons for school absence.[28] Speaking to a meeting in January 1949, one trade union leader described the general sense of depression that had settled over the workers at his firm. He indicated that many workers planned to drop their union membership if there were no additional distributions of vegetables and textiles.[29] But this instrumental pursuit of nourishment did not tell the whole story either. Both inside and outside of Germany, this material crisis shaped the locations from which people could contest the (symbolic) frameworks to more generally explain and define what was happening there.

Even as western propaganda played up the image of a "starving Soviet Zone,"[30] Berliners from all sectors continued to make forays into the surrounding areas to supplement their food supplies. Of course, people in the Soviet Zone *were* often hungry and regularly complained to officials there about shortages and inadequate and poor quality food distributions. The irony here was that in denouncing a "starving" zone, the western allies attributed a power to Soviet Zone authorities that they did not possess. Although there were real shortages, many farmers, shop owners, and private citizens simply took matters into their own hands and in the name of efficiency, profit, or greed, made black market arrangements to supplement what officials managed to allocate and control. Thus, Berliners coping with the restrictions of blockade were simply participants in a process that belied assertions and denunciations of control, which the great powers and their German supporters continued to make. In spite of border controls, these persons constantly developed new techniques to successfully transport goods into the city. Some used bicycles to circumvent train station checkpoints; others tossed goods from trains to accomplices waiting beside the tracks.[31] A wide variety of police reports hint at the broad scale of trade and transport across sector and zonal lines and suggest that police controls were hardly insurmountable. The control that the East desired and the West condemned simply did not exist.

These gaps did not, however, prevent occupation authorities from making new assertions of control, assertions which, while they did in fact exert substantial and at times destructive force, functioned in ways that often ran counter to the policies' supposed goals. On March 20, 1949, the Third Ordinance for Monetary Reform declared the Westmark to be the sole legal currency in Berlin's western sectors and dramatically transformed the basic economic environment in and around Berlin.[32] This currency directive, issued by the three western allies, did not make possession of Eastmarks illegal in the western sectors but did rule that Eastmarks need no longer be accepted at face value for such important transactions as food purchases or rent payment. Firms in the western sectors were now obligated to pay their employees entirely in Westmarks. The order established a wage clearance office in West Berlin to adjust the wages of those employed in the Soviet sector and drawing their wages in the western sectors or vice versa. This occurred at varying rates depending on the sector in which one worked, lived, and drew one's rations, and established a four-tier system of Westmark wage and salary payment. At the top tier were persons employed and drawing rations in the west sectors; they received one hundred percent of their wages in Westmarks. Those employed in the Soviet Sector but drawing their rations in the West were entitled to have sixty percent of their salary exchanged into Westmarks at a rate of 1:1. Employees working and drawing their rations in the Soviet Sector could exchange thirty percent of their wages into Westmarks at the rate of 1:1. Persons employed in the western sectors but drawing their rations in the East received only ten percent of their wages in Westmarks.[33] All other exchanges of currency had to be conducted at the going rate as set by private exchange Kiosks in the west sectors. The fact that all west sector obligations now had to be met in Westmarks, even while some Berliners continued to receive a portion of their income in Eastmarks, struck many as unfair.

German politicians in the western sectors had, of course, been pleading for such a step and greeted it with enthusiasm. Speaking to Berliners via radio, West Berlin mayor Ernst Reuter emphasized the political significance of the step: "Now they [the western allies] can not cut us off. . . . The simple fact of the introduction of the Westmark is worth more than a hundred political declarations [that] 'We'll stay here!' This is a fact—the Berliner desires facts and is completely justified in that."[34] For Reuter, the economic aspects of the currency decision were secondary. He had been looking for a decisive act demonstrating western commitment to the city; with the March 20 currency, he found it. The western sectors' economic foundation now lay decisively in the western zones. West Berlin had symbolically been extracted from its location in the midst of the Soviet Zone (and from the midst of its separate currency).

In this instance there was a significant disconnect between the assertions made about the material "reality" of this currency reform and the socio-cultural impact that it actually had. Just as the airlift claimed to ensure the material survival of West Berlin, the currency reform ostensibly sought to stabilize the economic situation in the city—at least this was its public face. But as Ernst Reuter emphasized, this was ultimately about locating West Berlin in the politico-economic realm of the emerging West German state. At the same time, while the East German authorities protested this "assault" on German unity, the policy greatly facilitated their ability to assert control over an unruly economic environment within their zone.

The communists sought to transform negative reactions to the currency reform into a broader political movement.[35] Speaking at a rally in the Soviet Sector, the Berlin communist party head denounced the currency reform as "nothing other . . . than a political provocation carried out with criminal swindler-methods. It is designed to tear Berlin totally apart and give new life to world-war-mongering by stirring up the Berlin conflict"[36] (Already by this time Berlin had become a symbolic expression to be deployed at multiple contested levels of meaning and political engagement.). There is little evidence that such efforts to expand the local political resonance of the currency reform were successful. Rather than looking to move against western policies, some workers implied that the communists should take additional steps to help these workers receiving only partial Westmark wages to meet their rent and other expenses.[37] Thus counter to the claims being made upon them by their party, these Berliners found a common advocacy across political and occupation lines, shaped by their common position and common experience of the negative impact of the currency reform.[38] But the conceptual ordering of that space was about to change again.

At midnight on 12 May, the provisions of the Soviet and American agreement to "end" the Blockade went into effect. Crossing gates went up on the Soviet Zone borders, and the first trucks set out with supplies for Berlin. The open flow of traffic between the western zones and Berlin began anew, and the Soviet Commander of Berlin, Alexander Kotikov, ordered the Berlin police to halt all searches on public transportation and remove police checkpoints on the sector borders. At the same time, he directed Police President Markgraf to prohibit the use of Westmarks in the Soviet Sector "under the strictest adherence to the pertinent regulations."[39] In line with this ongoing prohibition, the lifting of transportation restrictions did not imply a total refutation of existing control structures, and already on June 30, the Soviet Command in Berlin issued a directive to the police regarding "heightened controls" on the checkpoints that comprised the "Ring around Berlin." Although items such as hand luggage were to be allowed through unsearched, the desire to control traffic and transport in general remained.[40] The principles underlying Soviets' and German communists' desire for continued economic control were not being abandoned. Indeed, the expansion and consolidation of their control in the Soviet Zone and Soviet Sector of Berlin continued unabated.

In spite of increasingly concrete steps towards a state in which they would exercise almost total control, communist leaders kept looking over their shoulders—especially in Berlin. Speaking to the Berlin Central Committee on June 10, one party leader cautioned against assuming that the division of Berlin meant an end to the high points of conflict. He warned that a failure to put forth an all out effort would result in a "more dangerous situation" for the party and the entire Berlin working class.[41] In this he was simultaneously correct and mistaken. On a local stage, the communists could count on only limited popular support. They claimed to have the power to transform the city's dismal socioeconomic conditions but did not. At the same time, their position in power remained ultimately secure—because it did not depend on popular support but on Soviet military backing—but that did not overcome their perception of their own weakness. This uneasiness coalesced into real anxiety in the face of an extended violent clash that highlighted the symbolic and material geography of the conflict in/for Berlin: a June 1949 railroad workers' strike launched in response to the material impact of the March currency reform and in opposition to communist dominance of the central trade union organization. On some level, the internationally arranged resolution to the Blockade assuaged their uneasiness—the western powers at least tacitly accepted the emergence of a Stalinist East German state and the existence of a divided Berlin—but the railway workers strike and the resulting violence directly challenged any sense that a comfortable line had been drawn under that first Cold War clash.

Betwixt and Between: Berlin as Liminal Space

At midnight on May 20, the breakaway western trade union (UGO) declared a strike of railroad workers in Berlin.[42] A remarkable percentage of railroaders heeded the call, far more than the approximately 3,000 UGO members.[43] Over the course of the next few days, the strike evolved into a pitched battle for physical control of the train stations and rail infrastructure throughout Berlin's western sectors. These stations remained technically under the control of the Soviet Zone Railroad Authority and its railroad police, and West Berlin police officials had to accept these sites as extra-territorial exceptions to their authority, even though physically located in the midst of their jurisdictions. Of greatest interest here is not the fact that the strike turned violent (in spite of West Berlin police efforts at crowd control outside the stations) but rather the communists' perspective on their effort to "defend" *their* S-Bahn (commuter railway) in the middle of (West) Berlin. Rhetorically, they sought to integrate this effort into a broader conceptual assessment of the political situation in Berlin and Germany.

On one level, this appeal had a material component: the party called workers to aid the non-UGO/non-striking rail workers in keeping the S-Bahn running. One appeal denounced the strike as a threat both to the supply of Berlin and as targeted at the developing understanding between the great powers—their foreign ministers were set to meet in Paris.[44] Most obviously, communist leaders sought to deny that the walkout was a strike at all. The head of Berlin's communist-dominated central trade union organization criticized a discussion in the railway workers' union (not the breakaway union, but the official branch under the central organization's authority) in which the terms "strike" and "strikers" were used. He argued "that it is not about a strike at all; at most, it is an action. It could be very dangerous if even we introduce the ideology of a strike. We must characterize the action of the UGO [the breakaway union] and the western occupation powers as that which it is."[45] On one hand, the Berlin communists fought a battle over what this event should be called, a first effort to transform the stance from which their supporters could approach it.[46] Yet they also moved to occupy the material locations at the center of the strike.

For little more than a day after the strike declaration, railway officials managed to keep the commuter railroad (S-Bahn) running in Berlin's western sectors, but by the morning of May 22, striking workers began to gain the upper hand. Shortly after the initial strike reports reached party officials, further information came in about the situation at the Charlottenburg and Zoological Garden stations, located in the British Sector. The railroad authority used the S-Bahn to ferry reinforcements of Railroad police and communist party and youth group members to emerging hot spots, and they took up positions in the stations and along the tracks. At the Zoo station:

> Rowdies urged on by the UGO throw stones, pieces of the railing, and iron rails from the bridge onto the railway embankment. The Railroad police guarding a switch box about 300m away begin to shoot so that the bullets fly just over the bridge. On the bridge, some comrades have spread out and started numerous discussion groups. Discussion leadership excellent. [The west sector police] intervenes in the face of a hail of stones and brings it to a stop. People relatively quiet, somewhat disturbed by the shooting, [west sector police] extremely disciplined in order to prevent any provocation.[47]

The tension between the violence and communist efforts to form discussion groups highlights party members' uneasy presumption that they could control—if not the crowds themselves—at least the meaning attributed to or appropriated by those crowds. More significantly this description elaborates the ebb and flow of violence in between the official authorities of East and West—to some extent regulated by those authorities but certainly not under their control. It describes a kind of "liminal space" that is at once powerful and unsettling, that not only suggests possible explanatory meanings (to the communists as well as to other actors in Berlin), but also hints that it might undermine the very structures that such explanations presume to support.[48]

As the strike dragged on into June, it posed an increasing dilemma for western officials. The Paris Council of Foreign Ministers began its meetings on May 23 to discuss the situation in Berlin and in Germany—a conference convened as part of the agreement that "ended" the Berlin Blockade. At the same time, western officials faced the unpleasant reality that the renewed transportation difficulties (at least with respect to rail traffic) caused by the Berlin strike lent at least some credibility to communist claims of a western "self-blockade." For the most part, western politicians and military officials pursued a policy of ongoing de-escalation, but the strikers continued to hold out. The communists, on the other hand, sought to utilize the strike as a propaganda windfall. Its leaders portrayed the "UGO-putsch" as part of a broad western plan to counter the evolving state structures in the Soviet Zone and disrupt the foreign ministers' discussions in Paris. In an introduction to the brochure provided by the East Berlin city council for delegates arriving in the city for a congress that would lay the political foundations of East Germany, the East Berlin mayor decried the west sector politicians who strove "even with the assistance of vandalistic acts of destruction, designed to destroy the relaxation of the international situation brought about by the peace politics of the USSR, because they fear a normalization of life in Berlin."[49] This conflation of local and international issues in communist assessments of the situation is telling. For party leaders, the "S-Bahn provocation" or "UGO-putsch" became a catch phrase that sought to demonstrate the extent to which they, too, were targets of western "assaults" on "progressive" forces around the world.[50] Simply the fact that communist party officials in Berlin exerted so much effort to mobilize party members and the party youth groups to occupy train stations and "defend" them against striking workers suggests that they viewed the strike as much more than a labor dispute.[51] Indeed, this act of "defense" was, for the communists, at least as important as the specific demands of the strikers. But it also pointed to the degree to which these officials presumed that these diverse locations could be incorporated into a single framework of (symbolic) meaning.

In the last week of June, the S-Bahn strike still continued. Strikers had overwhelmingly rejected a settlement offer that called for the railway authority to pay workers living in the western sectors sixty percent of their wages in Westmarks. Over the course of the strike, UGO membership had increased significantly, and the striking railroaders seemed poised to wait for acceptance of all of their demands.[52] Finally, the western occupiers intervened. In a 25 June directive, they ordered the West Berlin city council to end financial support to striking workers and adopt a compromise solution on the exchange of railroader wages. The strike came to a close, although train traffic was not fully reestablished until 1 July. To some extent, the strike's end allowed an "honorable defeat" for UGO leaders, enabling them to establish solid credentials as powerful political force in (West) Berlin.[53] On the other hand, the fact that a strike against the Soviet-sponsored railway authority came to an end as a result of a direct order from the western allies produced some tension between the three western powers and the "most militantly democratic forces" in the ranks of Berlin workers.[54] At the very least, the agreement imposed a preliminary solution to the problem of Westmark payment for railroad employees. However, the outcome of the S-Bahn strike also sheds broader light on the conclusion of the blockade crisis as a whole. It points to the way in which events and actors "on the ground" drove the escalation of tensions in the city. Ultimately, the Cold War stability of an iconic location like Berlin depended upon a collaborative undertaking (even if it was an unintentional collusion) of the great powers, East and West. Yet the tension, which made this location significant in and for the Cold War depended on local participation that these Great Power actors neither fully understood nor could fully control.

The outcome of this crisis—the parallel resolution of great power decision-making and the enforced if temporary settlement of the messy, material struggle on the ground in Berlin—points to a suggestive new way to consider the role of location and symbolic construction in forming the Cold War. Indeed, I would like to argue that the tension between these two arenas is precisely what *constitutes* the Cold War—for the first time—in mid-1949. This assertion proposes that we consider the Cold War to be a particular style of conflict, one imbedded in twentieth century mass politics; a conflict for mass consumption; and one fought out according to evolving ritual practices. Acting in situations that remained incomprehensible and messy, the great powers facilitated the production of ritual commitments to a conflict that was defined primarily by its disconnect from the complicated and difficult conditions on the ground. Ironically, this messiness and the relative freedom for local individuals to act on their own behalf exerted pressure on the symbolic framework (which they also sought to exploit) and tightened its explanatory hold on a battleground for which its actual descriptive power was actually quite minimal.

Conclusion

Postwar Berlin encompassed multiple layers of betwixt and between-ness: occupied Germany—a divided state extracted from the normal international arena; Berlin as a separately defined space within the Soviet Zone of occupation; the S-Bahn and its stations as extraterritorial bastions in the midst of the city's western sectors—these political definitions were sublated (*aufgehoben*) in the material encounter that took place within and across these presumed structures as Berliners struggle to survive in the midst of material devastation. But Berlin's liminality functioned even more powerfully in the symbolic encounter of these microscopic structures with a macroscopic (global) framework that remained both enmeshed in and distinct from each other. Only by enframing the location of their encounter—a site that both undermined and facilitated their respective projects of meaning production—can we elaborate the symbolic violence inherent to the Cold War. Ultimately, this process points to the power of a symbolic structure (the Cold War) to assume truly global dimensions—even if this success depended on a liminal encounter that remained both unrecognized and (certainly in policy terms) unaddressed. While the tension implicit in this excluded social space (Turner's antistructure) lent the Cold War its potency, it revolutionary potential also threatened to send the whole structure careening out of control. Thus, as a warning about the power of the liminal encounter, this also suggests—in a time where true believers are again being called into service for new global conflicts—that the functional legacy of the Cold War is not yet history.

Notes

1. For a brief overview of the history of everyday life and its critics, see Georg Iggers, *Historiography in the Twentieth Century: From Scientific Objectivity to the Postmodern Challenge* (Hanover and London: Wesleyan University Press, 1997), 101-17.

2. In English, there are at least two books with the title, *Battleground Berlin*. Ruth Andreas-Friedrich's memoir, *Battleground Berlin: Diaries 1945-1948*, trans. Anna Boerresen (New York: Paragon House Publishers, 1990), whose title is a poor translation of the German original, *Schauplatz Berlin* [perhaps better, Scene: Berlin]. Murphy and Kondrashev's account of Cold War espionage in Berlin is a more typical rendering of great power agents (literally) facing off in the city. See David E. Murphy and Sergei A. Kondrashev, *Battleground Berlin: CIA vs. KGB in the Cold War* (New Haven, Conn.: Yale University Press, 1999).

3. See "Text of Kennedy Statements in Berlin," *The New York Times*, 27 June 1963, 12.

4. Richard Harker, Cheleen Mahar, and Chris Wilkes, eds., *An Introduction to the Work of Pierre Bourdieu: The Practice of Theory* (London: Macmillan, 1990), 5. See also the introduction to Murray Edelman, *The Symbolic Uses of Politics* (Urbana: University of Illinois Press, 1974).

5. Pierre Bourdieu, *Die verborgenen Mechanismen der Macht*, ed. Margarete Steinrücke, trans. Jürgen Bolder et al, Schriften zu Politik und Kultur 1 (Hamburg: VSA-Verlag, 1992), 174.

6. Pierre Bourdieu, *Praktische Vernunft: Zur Theorie des Handelns*, trans. Hella Beister (Frankfurt am Main: Suhrkamp, 1998), 116.

7. Harker, et al, *An Introduction*, 16.

8. Bourdieu, *Praktische Vernunft*, 164ff. Bourdieu provides an example of the principle of *do et des* [I give, so that you give] in which a gift is given in expectation of another being returned. All the participants know that, but act as if that was not what they did. In the same way, when we give a gift, we remove the price tag, in a way denying the reality of the power implicit in this gift-granting authority.

9. Here I depart somewhat from Murray Edelman's assessment of political symbols. For him, they serve to conceal the actual functioning of material politics behind its alleged mobilization on behalf of popular interest; politics itself becomes largely a ritual process, a rite meant to reassure the public but in no way driven by its desires. Lost in this description is a clear sense of agency. If political leaders are not acting on behalf of the popular interest but serve essentially as dueling bureaucrats who haggle over the accumulation and distribution of material benefits, then who produces the symbolic structures that explain their actions? See Murray Edelman, *Politics as Symbolic Action: Mass*

Arousal and Quiescence, Institute for research on poverty monograph series (Chicago: Markham Publishing Co., 1971), 159-60.

10. My point here draws significantly on Bourdieu's notion of the *habitus*. He defines habitus as "a system of durable, transportable dispositions which functions as the generative basis of structured, objectively unified practices." Pierre Bourdieu, *Algeria 1960: The Disenchantment of the World, the Sense of Honour, the Kabyle House, or the World Renewed* (Cambridge: Cambridge University Press, 1979), vii, quoted in Harker, et al, *An Introduction*, 10. The habitus does not prescribe a particular action; rather it marks the boundaries of those actions which a particular actor would find at all conceivable. This description serves to emphasize that the elaboration of symbolic systems (of meaning) happens not from the top-down (per a set of rules) but on the bases of the strategies individuals adopt with respect to the positions they have assumed. Bourdieu, *Die verborgenen Mechanismen*, 33 and Harker, et al, *An Introduction*, 202-3.

11. Turner describes liminality as follows: "A limen is a threshold, but at least in the case of protracted initiation rites or major seasonal festivals it is a very long threshold, a corridor almost, or a tunnel which may become a pilgrim's road. . . . Let us refer to the state and process of midtransition as 'liminality.'. . . Those undergoing it . . . are betwixt and between established states of politico-jural structure. They evade ordinary cognitive classification, too, for they are not this or that, here or there, one thing or the other." Victor Turner, *Blazing the Trail: Way Marks in the Exploration of Symbols*, ed. Edith Turner (Tucson and London: University of Arizona Press, 1992), p 49. For a general discussion of this liminal condition, see Victor Turner, "Betwixt and between: the liminal period in *rites de passage*," in *The Forest of Symbols: Aspects of Ndembu Ritual* (Ithaca, N.Y., and London: Cornell University Press, 1973) 93-110.

12. Victor Turner, *Blazing the Trail*, 133.

13. Turner, 153.

14. Turner, 153.

15. In this regard, Pierre Bourdieu notes: "The only thing that separates [the pope, a president, or a general secretary] from a character in a comedy or from a madman with delusions of grandeur (*Größenwahnsinniger*) is that in general that people take him seriously." Bourdieu, *Die verborgenen Mechanismen*, 86.

16. For example, in his announcement of an arms reduction treaty to be signed with Russia, American President George W. Bush announced on May 13, 2002, that "[t]his treaty will liquidate the legacy of the Cold War." "Bush Announces U.S.-Russia Nuclear Arms Reduction Agreement," from *U.S. Department of State: International Information Programs* May 13, 2002, http://usinfo.state.gov/topical/pol/arms/02051310.htm, accessed July 28, 2002.

17. Most tellingly, John Lewis Gaddis, *The Long Peace: Inquiries into the History of the Cold War* (New York and London: Oxford University Press, 1989).

18. Guy Oakes, *The Imaginary War: Civil Defense and American Cold War Culture* (New York and Oxford: Oxford University Press, 1994), 9.

19. Ironically, one of the reviewers of Oakes's book commends it for its insight into the "politics of everyday life" of the Cold War. While the book is an outstanding critique of the political production and manipulation of culture, it is precisely in the sense in which culture remains an object of policy that the book does *not* interject broader everyday agency into the mix. See, JoAnne Brown, "Guy Oakes, The Imaginary War: Civil Defense and American Cold War Culture," review of *The Imaginary War: Civil Defense and American Cold War Culture* by Guy Oakes, *The Journal of American History* 83, no. 4 (March 1997), 1481. Another author makes a broader argument about the Cold War itself as the imaginary product of domestic political demands in both the United States

and Soviet Union. See Mary Kaldor, *The Imaginary War: Understanding the East-West Conflict* (Oxford and Cambridge, Mass.: Blackwell, 1990).

20. It is precisely this regard that one can understand the long-running academic dispute about the origins of the Cold War. Whether "traditional" (Soviet/communist expansionism), "revisionist" (American imperialism), or "post-revisionist" (parallel miscalculations/domestic pressures), these trends in American scholarship all seek to uncover the true motivations for the conflict. For a good summary of the progression of these debates, see Anders Stephanson's chapter on the United States and the Cold War in *The Origins of the Cold War in Europe: International Perspectives*, ed. David Reynolds (New Haven, Conn. and London: Yale University Press, 1994), 23-52.

21. I appropriate Gertrude Stein's famous remark about Oakland, Calif.

22. Volker Koop: "Die perfekte Blockade? Über Mythos und Legendenbildung," in *Auftrag Luftbrücke: Der Himmel über Berlin 1948-1949*, ed. Uwe Förster, Stephanie v. Hochberg, Ulrich Kubisch and Dietrich Kuhlgatz (Berlin: Nicolai, 1998), 361-369 and Volker Koop, *Kein Kampf um Berlin? Deutsche Politik zur Zeit der Berlin-Blockade 1948/1949* (Bonn: Bouvier Verlag, 1998). Gerhard Keiderling, *"Rosinenbomber" über Berlin: Währungsreform, Blockade, Luftbrücke, Teilung* (Berlin: Dietz Verlag, 1998). William Stivers, "The Incomplete Blockade: Soviet Zone Supply of West Berlin, 1948-49," *Diplomatic History* 21, no. 4 (Fall 1997), 569-602.

23. This is a central argument in my dissertation, "More than an airlift: constructing the Berlin Blockade as a Cold War battle, 1946-1949" (Ph.D. diss., University of Chicago, 1999).

24. In their concluding critique of the work of Pierre Bourdieu, a group of New Zealand scholars assert that it is particularly in the link between the material and symbolic that one encounters "social reality." In Harker, et al, *An Introduction*, 196.

25. See e.g., Ann and John Tusa, *The Berlin Airlift* (New York: Athenaeum, 1988), 263, 282.

26. On this phenomenon, see the detailed account in William Stivers, "The Incomplete Blockade," 584ff.

27. See Jeffrey Kopstein, *The Politics of Economic Decline in East Germany, 1945-1989* (Chapel Hill and London: University of North Carolina Press, 1997), 21.

28. Hauptamt Statistik: Erhebung über die sozialen Verhaeltnisse der Berliner Schulkinder, 1947/48, Landesarchiv Berlin (LAB), STA Rep. 120, Nr. 3303.

29. Protokoll of the Bezirks-Ausschuss Tiergarten on Jan 11, 1949; report dated Jan. 20, 1949, LAB, FDGB 136.

30. See William Stivers, "The Incomplete Blockade," 577.

31. For a discussion of these and other techniques, see the weekly communiqués produced by the Brandenburg police, Br. LHA, Ld. Br. Rep. 203, Nr. 117.

32. For the full text of the order, see LAB, Rep 280 (LAZ-Sammlung), Sta.-Ort.-Nr. 1283. From the end of June 1948 until the issuing of this order, both East- and Westmarks circulated as legal currency in Berlin's western sectors.

33. See the Third Ordnance for Monetary Reform (Supplemental Currency Ordinance), LAB, Rep. 280 (LAZ-Sammlung), Sta.-Ort Nr. 1283. See also Gerhard Keiderling, "*Rosinenbomber*," 183ff.

34. Willy Brandt and Richard Löwenthal, *Ernst Reuter, ein Leben für die Freiheit: eine politische Biographie* (Munich: Kindler Verlag, 1957), 501. According to Brandt and Löwenthal, Reuter also greeted the publication of the text of the North Atlantic treaty [two days earlier] as an indication that (West) Berlin was now firmly entrenched in the ranks of Western Europe.

35. Technically, there was no communist party in Berlin. In April 1946, the communist and social democratic parties in the Soviet Zone joined to become the Socialist Unity Party of Germany (SED). This party increasingly fell under the Stalinist line of former communists, who increasingly marginalized and persecuted former socialists. Berlin remained the only city in the Soviet Zone where, as a result of its four-power status, the socialists maintained a separate existence. For convenience, in this paper I will simply refer to the SED as "communist."

36. "Berlin gegen den Währungsbetrug: 500000 demonstrieren gegen Kriegshetze und Währungsgeschwindel—Hans Jendretzkys Aufruf zur Selbtshilfe—Es gibt keine 'Grenzgänger' in Berlin!" *Tägliche Rundschau* (March 27, 1949), 1.

37. Bericht über Währungsdiskussionen in Betrieben der Westsektoren, dated March 21, 1949, LAB, BPA, FDGB 1557.

38. Bourdieu would explain this tendency toward group formation in terms of the proximity of individuals within a social space along with a concomitant habitus. See Bourdieu, *Die praktische Vernunft*, 50.

39. Kotikov to Markgraf, LAB, STA Rep. 26, Nr. 15, Bl. 86. The German translation of the directive is found on the preceding page in the file.

40. Auszug aus der Instruktion über Ordnung des Durchlassens der Personen und Ladungen durch die Kontrolldurchlassungspunkte auf der äusseren Grenze von Groß-Berlin ab 12. Mai 1949, LAB, STA Rep. 26, Nr. 143, Bl. 94. A west sector report described how Soviet Zone police did in fact confiscate foodstuffs and other goods from travelers' luggage in early July. See "Zugkontrollen und Lebensmitteldiebstähle: Russisches Interesse für Fischkonserven—Bewaffnete Posten an den Ausfallstraßen." *Tagesspiegel* (July 6, 1948), in BA-Lichterfelde, DM 1/3135.

41. Protokoll d. Landesvorstandssitzung, June 10, 1949, LAB, BPA, IV L-2/1/45, Bl. 91ff.

42. The Unabhängige Gewerkschaftsorganisation (Independent Trade Union Organization) had previously functioned as an opposition group within the central trade union federation in Berlin.

43. Arthur Schlegelmilch cites an OMGBS report that put the number of strikers at 10,900, those in doubt at 3000, and the number of "strike breakers" at 1000. In Schlegelmilch, *Hauptstadt im Zonendeutschland: Die Entstehung der Berliner Nachkriegsdemokratie 1945-1949* (Berlin: Haude & Spener, 1993), 247, n. 402.

44. Memo in the subsection Mitte; addressed to all BA re. impending S-Bahn strike, dated May 20, 1949, LAB, BPA, FDGB 136.

45. Bericht über die Besprechung am May 30, 1949 [IG Eisenbahn?], LAB, BPA, FDGB 84, Bl. 150.

46. Pierre Bourdieu notes, "In politics nothing is more realistic that the battle over words. To place one word in the place of another, means to transform the view of the social world and thereby to contribute to its transformation." Bourdieu, *Die verborgenen Mechanismen*, 84.

47. Meldung Niki Sternberg—Karl Morgenstern (13.45); no date but probably 22.5.1949, LAB, BPA, IV L-2/10/402.

48. On the threat implicit in liminal spaces, see Victor Turner, *Blazing the Trail*, 153.

49. Brochure (1949) in Nachlaß Wilhelm Pieck, SAPMO, NY 4036/742. The Third People's Congress was held in Berlin from May 29 to June 3, 1949. Its central act was the adoption of a constitution for a "German Democratic Republic."

50. It is interesting to note the extent to which SED leaders regularly refer to Communist advances in China as forcing the western hand in Europe and Germany. See e.g., the

remarks of Otto Grotewohl, Stenographisches Bericht der 20. (34.) Parteivorstandssitzung der SED, July 20-21, 1949, SAPMO, DY 30/IV 2/1/67, Bl. 21.

51. In this regard, see e.g., the numerous reports from district SED groups on their activities during the strike in LAB, BPA IV L-2/10/402.

52. Schlegelmilch, *Hauptstadt im Zonendeutschland*, 246ff. See also Gerhard Keiderling, "*Rosinenbomber*," 299-300.

53. Schlegelmilch, *Hauptstadt im Zonendeutschland*, 249-50. Also Brandt and Löwenthal, *Ernst Reuter*, 527f.

54. Brandt and Löwenthal, *Ernst Reuter*, 527.

Selected Bibliography

Alexander, Christopher. *Notes on the Synthesis of Form*. Cambridge, Mass.: Harvard University Press, 1964.

Allison, Robert J. *The Crescent Obscured: The United States and the Muslim World, 1776-1815*. New York: Oxford University Press, 1995.

Anderson, Perry. *Passages from Antiquity to Feudalism*. London: New Left Books, 1974.

———. *Lineages of the Absolutist State*. London: New Left Books, 1974.

Andreas-Friedrich, Ruth. *Battleground Berlin: Diaries 1945-1948*. Translated by Anna Boerresen. New York: Paragon House Publishers, 1990.

Aristotle. *Physics*. Translated by Hippocrates G. Apostle. Grinnell, Iowa: Peripatetic Press, 1980.

Arrighi, Giovanni. *The Long Twentieth Century: Money, Power, and the Origins of Our Times*. London and New York: Verso, 1994.

Backhaus, Gary and John Murungi, eds. *Transformations of Urban and Suburban Landscapes: Perspectives from Philosophy, Geography, and Architecture*. Lanham, Md.: Lexington Books, 2002.

Baudrillard, Jean. *Pour une critique de l'économie politique du signe*. Paris: Gallimard, 1972.

Bennett, Brandon. "What is a Forest ? On the Vagueness of Certain Geographic Concepts." *Topoi* 20 (2001): 189-201.

Berque, Jacques. "Vie sociale et variations de mode et de densité." *Sociologie des mutations*. Edited by Georges Balandier. Paris: Éditions Anthropos, 1970.

Bilsky, Lester J., ed. *Historical Ecology: Essays on the Environment and Social Change*. Port Washington, N.Y.: Kennikat Press, 1980.

Blaikie, Piers, and Harold Brookfield. *Land Degradation and Society*. London: Methuen, 1987.

Borgo, Stafano, Nicola Guarino, and Claudio Masolo. "Stratified Ontologies: The Case of Physical Objects." *European Conf. on AI (ECAI96)*. Workshop on Ontological Engineering, Budapest, Hungary, August 1996.

———. *Towards An Ontological Theory of Physical Objects*. National Research Council, LADSEB-CNR, Italy, 1996.

Bourdieu, Pierre. *Algeria 1960: the Disenchantment of the World, the Sense of Honour, the Kabyle House, or the World Renewed*. Cambridge: Cambridge University Press, 1979.

———. *Praktische Vernunft: Zur Theorie des Handelns*. Translated by Hella Beister. Frankfurt am Main: Suhrkamp, 1998.

———. *Die verborgenen Mechanismen der Macht*. Edited by Margarete Steinrücke. Translated by Jürgen Bolder et al, Schriften zu Politik und Kultur 1. Hamburg: VSA-Verlag, 1992.

Boyd, Julian P. ed. *The Papers of Thomas Jefferson*. Princeton, N.J.: Princeton University Press, 1930-2003.

Brandt, Willy and Richard Löwenthal. *Ernst Reuter, ein Leben für die Freiheit: eine politische Biographie*. Munich: Kindler Verlag, 1957.

Braudel, Fernand. *Afterthoughts on Material Civilization and Capitalism*. P.M. Ranum, trans. Baltimore, Md.: The Johns Hopkins University Press, 1977.

———. *Le Mediterranee et le monde Mediterraneen a l'epoque de Philippe II*. 2e ed. revue et augmente. 2 vol. Paris: Lib. Armand Colin, 1966.

———. *The Mediterranean and the Mediterranean world in the age of Philip II*. 2 vols. S. Reynolds, trans. New York: Harper and Row, 1972-1973.

———. *The Perspective of the World*. Sian Reynolds, trans. New York: Harper and Row, 1984.

———. *The Structures of Everyday Life*. Sian Reynolds, trans. New York: Harper and Row, 1981.

———. *The Wheels of Commerce*. Sian Reynolds, trans. New York: Harper and Row, 1982.

Brenner, Neil. "Beyond state-centrism? Space, territoriality, and geographical scale in globalization studies." *Theory and Society* 28 (1999), 39-78.

Brenner, Robert. "Agrarian Class Structure and Economic Development in Pre-Industrial Europe." In *The Brenner Debate*, edited by T. H. Aston and C. H. E Philpin. Cambridge: Cambridge University Press, 1985, 10-62.

Brockway, L. *Science and Colonial Expansion: The Role of the British Botanic Gardens*. New York: Academic Press, 1979.

Broich, John. "The Wasting of Wolin: Environmental Factors in the Downfall of a Baltic Town." *Environment and History* 7 (2001): 187-199.

Brooke, John L. "Ancient Lodges and Self-Created Societies: Voluntary Association and the Public Sphere in the Early Republic." In *Launching the "Extended Republic": The Federalist Era*, Ronald Hoffman and Peter J. Albert, Charlottesville: University Press of Virginia, 1996, 273-377.

Brown, Richard D. *Knowledge is Power: The Diffusion of Information in Early America, 1700-1865*. New York: Oxford University Press, 1989.

———. *The Strength of a People: The Idea of an Informed Citizenry in America, 1650-1870*. Chapel Hill: University of North Carolina Press, 1996.

Bunker, Stephen G. *Underdeveloping the Amazon*. Urbana: University of Illinois Press, 1985.

Burkett, Paul. *Marx and Nature: A Red and Green Perspective*. New York: St. Martin's Press, 1999.

Burnett, Ron. "Critical Approaches to Culture." *Communications + Hypermedia*. http://www.eciad.bc.ca/~rburnett/weekseven.html.

Bushev, Michael. *Synergetics: Chaos, Order, Self-Organization*. Singapore: World Scientific, 1994.

Buttimer, Anne. *Geography and the Human Spirit*. Baltimore: The Johns Hopkins University Press, 1993.

Calhoun, Craig, ed. *Habermas and the Public Sphere*. Cambridge, Mass.: MIT Press, 1992.

Cantwell Smith, Brian. *On the Origin of Objects*. Boston: The MIT Press, 1996.

Carey, Mathew. *A Short History of Algiers*. Philadelphia: Carey, 1794.

Casati, Roberto, and Achile C. Varzi. *Holes and Other Superficialities*. Boston: MIT Press, 1994.

———. "Ontological Tools for Space and Geographic Representation." *Workshop on "Formal Models of Common-Sense Geographic Worlds."* San Marcos, Tex., October 30, 1996.

———. "Spatial Entities." *Spatial and Temporal Reasoning*. Edited by Oliviero Stock. Dordrecht: Kluwer 1997.

Casey, Edward S. *Getting Back into Place: Toward a Renewed Understanding of the Place-World*. Bloomington: Indiana University Press, 1993.

Chaunu, Pierre. "Bresil et l'Atlantique au XVIIe Siecle." *Annales E.S.C.* 17 (6) (1961): 1176-1207.

Chew, Sing. *World Ecological Degradation*. Walnut Creek, Calif.: Alta Mira Press, 2001.

Ciccantell, Paul, and Stephen Bunker. "Economic Ascent and the Global Environment: World-Systems Theory and the New Historical Materialism." In *Ecology and the World-System*, edited by Walter L. Goldfrank, Andrew Szasz, and David Goodman. Westport, Conn.: Greenwood Academic Press, 1999, 107-122.

Clark, Charles E. *The Public Prints: The Newspaper in Anglo-American Culture, 1665-1740*. New York: Oxford University Press, 1994.

Clemens, Samuel Langhorne. *Adventures of Huckleberry Finn*. New York: W. W. Norton, 1961.

Cohen, R.S. ed. *Inquiries and Provocations, Selected Writings of Herbert Feigl 1929-1974*. Dordrecht and London: D. Reidel Publishing Company, 1981.

Cranston, Maurice. *Jean-Jacques: The Early Life and Work of Jean-Jacques Rousseau, 1712-1754*. Chicago: University of Chicago Press, 1982.

———. *The Noble Savage: Jean-Jacques Rousseau, 1754-1762*. Chicago: University of Chicago Press, 1991.

———. *The Solitary Self: Jean-Jacques Rousseau in Exile and Adversity*. Chicago: The University of Chicago Press, 1997.

Cronon, William. *Nature's Metropolis: Chicago and the Great West*. New York: W.W. Norton, 1991.

Crosby, Alfred W. *The Columbian Exchange: Biological and Cultural Consequences of 1492*. Westport, Conn.: Greenwood Academic Press, 1972.

———. *Ecological Imperialism: The Biological Expansion of Europe, 900-1900*. Cambridge: Cambridge University Press, 1986.

Dant, Tim. *Material Culture in the Social World*. London: Open University Press, 1999.

Davidson, Cathy. *Revolution and the Word: The Rise of the Novel in America*. New York: Oxford University Press, 1986.

Davis, Mike. *Late Victorian Holocausts: El Nino Famines and the Making of the Third World*. London: Verso, 2001.

de Shalit, Avner. "Ruralism or Environmentalism?" *Environmental Values* 5 (1996): 47-58.

Diamond, Jared. *Guns, Germans, and Steel: The Fates of Human Societies*. New York: W. W. Norton, 1997.

Dippert, Randall R. *Artifacts, Arworks, and Agency*. Philadelphia: Temple University Press, 1993.

Doerflinger, Thomas M. *A Vigorous Spirit of Enterprise: Merchants and Economic Development in Revolutionary Philadelphia*. Chapel Hill: University of North Carolina Press, 1986.

Drayton, Richard. *Nature's Government: Science, Imperial Britain, and the "Improvement" of the World*. New Haven, Conn.: Yale, 2001.

Dunaway, Wilma A. "Incorporation of Mountain Ecosystems into the Capitalist World-System." *Review* 19 (4) (1996): 355-381.

Eco, Umberto. "Interpretation and History." In *Interpretation and Overinterpretation*, edited by Stefan Collini. New York: Cambridge University Press, 1992.

Edelman, Murray. *The Symbolic Uses of Politics.* Urbana: University of Illinois Press, 1974.

———. *Politics as Symbolic Action: Mass Arousal and Quiescence.* Institute for Research on Poverty Monograph Series. Chicago: Markham Publishing Co., 1971.

Elvin, Mark, and Liu Ts'ui-jung, eds. *Sediments of Time: Environment and Society in Chinese History.* Cambridge: Cambridge University Press, 1998.

Fagan, Brian M. *Floods, Famines, and Emperors: El Nino and the Fate of Civilizations.* New York: Basic Books, 1999.

Feigl, Herbert. *Inquiries and Provocations, Selected Writings, 1929-1974.* Edited by R. S. Cohen. Dordrecht: D. Reidel Publishing Company, 1981.

Ferry, Luc. *The New Ecological Order.* Translated by Carol Volk. Chicago: University of Chicago Press, 1995.

Foster, John Bellamy. *The Vulnerable Planet: A Short Economic History of the Environment.* New York: Monthly Review Press, 1994.

———. *Marx's Ecology.* New York, Monthly Review Press, 2000.

Friedmann, Harriet. "What on Earth is the Modern World-System? Foodgetting and Territory in the Modern Era and Beyond." *Journal of World-Systems Research* 6 (2) (Summer, 2000): 480-515.

Fuller, Richard Buckminster. *Cosmography: A Posthumous Scenario for the Future of Humanity.* Adjuvant, Kiyoshi Kuromiya. New York : Macmillan Publishing Company, 1992.

Fuller, Richard Buckmister, in collaboration with E. J. Applewhite. *Synergetics: Explorations in the Geometry of Thinking.* New York: Macmillan Publishing Co., 1975.

———. *Synergetics 2. Further Explorations in the Geometry of Thinking.* New York: Macmillan Publishing Co., 1979.

Gaddis, John Lewis. *The Long Peace: Inquiries into the History of the Cold War.* New York and London: Oxford University Press, 1989.

Gadgil, Madhav, and Ramachandra Guha. *This Fissured Land: An Ecological History of India.* Berkeley: University of California Press, 1992.

Galison, Peter and David J. Stump, eds. *The Disunity of Science: Boundaries, Contexts and Power.* Stanford, Calif.: Stanford University Press, 1996.

Gärling, T., A. Böök, and E. Lindberg. "The Acquisition and Use of an Internal Representation of the Spatial Layout of the Environment During Locomotion." *Man-Environment Systems* 9 (1979): 200.

Giere, Ronald. *Explaining Science: A Cognitive Approach.* Chicago: Chicago University Press, 1988.

———. *Science without Laws.* Chicago: The University of Chicago Press, 1999.

Gieryn, Thomas F. *Cultural Boundaries of Science.* Chicago: The University of Chicago Press, 1999.

Glendening, Parris N. "What is Smart Growth?: A Message from Governor Glendening." http://www.smartgrowth.state.md.us/whatis.htm.

Goffman, Erving. *Frame Analysis: An Essay on the Organization of Experience.* Boston: Northeastern University Press, 1986.

Golledge, Reginald G. and Robert J. Simpson, eds. *Spatial Behavior: A Geographic Perspective.* London: Guilford, 1997.

Gopnik, Alison. "Scientist as Child." *Philosophy of Science* 63, no. 4 (Dec. 1997): 485-514.

Grant, Michael. *History of Rome.* New York: Charles Scriber's Sons, 1978.

Grew, Raymond, editor. *Food in Global History*. Boulder, Colo.: Westview, 2000.

Griffiths, Tom, and Libby Robin, eds. *Ecology and Empire: Environmental History of Settler Societies*. Seattle: University of Washington Press, 1997.

Grove, Richard H. *Ecology, Climate, and Empire: Colonialism and Global Environmental History, 1400-1940*. Cambridge: White Horse Press, 1997.

——— *Green Imperialism: Colonial Expansion, Tropical Island Edens, and the Origins of Environmentalism, 1600-1860*. Cambridge: Cambridge University Press, 1995.

Grove, Richard H., and John Chappell, eds. *El Nino, History, and Crisis: Studies from the Asia-Pacific Region*. Cambridge: White Horse Press, 2000.

Grove, Richard H., Vinita Damodaran, and Satpal Sangwan, editors. *Nature and the Orient: The Environmental History of South and Southeast Asia*. New Delhi: Oxford University Press, 1998.

Guarino, Nicola, and Christopher Welty. "Identity, Unity, and Individuality: towards a Formal Toolkit for Ontological Analysis." *ECAI-2000: Proceedings of the 14th European Conference on Artificial Intelligence*, edited by Werner Horn. Berlin: IOS Press: 219-23.

Gunn, Alistair S. "Rethinking Communities: Environmental Ethics in an Urbanized World." *Environmental Ethics* 20 (1998): 341-360.

Habermas, Jurgen. *The Structural Transformation of the Public Sphere: An Inquiry into a Category of Bourgeois Society*. Cambridge, Mass.: MIT Press, 1989.

Haila, Yrjo, and Richard Levins. *Humanity and nature: Ecology, science, and society*. London: Pluto, 1992.

Hamilton, Edith. *The Greek Way*. New York: W. W. Norton & Co, 1983.

Harker, Richard, Cheleen Mahar, and Chris Wilkes, eds. *An Introduction to the Work of Pierre Bourdieu: the Practice of Theory*. London: Macmillan, 1990.

Hartshorne, Richard. *The Nature of Geography: A Critical Survey of Current Thought in the Light of the Past, Annals of the Association of American Geographers*, Volume XXIX, Numbers 3 and 4. Lancaster, Penn.: Association of American Geographers, 1939.

Harvey, David. *Spaces of Hope*. Berkeley: University of California Press, 2000.

———. *The Limits to Capital*. London: Verso, 1999.

Heath, Louise Robinson. *The Concept of Time*. Chicago: University of Chicago Press, 1936.

Hegel, G. W. F. *Philosophy of Mind*. Translated by William Wallace. Oxford: Clarendon Press, 1971.

Heidegger, Martin. *Being and Time*. Translated by John Macquarrie and Edward Robinson. New York: Harper & Row, 1962.

Helleiner, Karl. "The Population of Europe from the Black Death to the Eve of the Vital Revolution." In *The Cambridge Economic History of Europe: The Economy of Expanding Europe in the Sixteenth and Seventeenth Centuries*, vol. IV, edited by E. E Rich and C. H. Wilson. London: Cambridge University Press, 1967, 1-95.

Herodotus. *The Histories*. Translated by Aubrey de Selincourt. Baltimore: Penguin Books, 1954.

Heschel, Araham Joshua. *The Sabbath*. New York: Noonday Press, 1975.

Hilpinen, Risto. "Artifacts." *Stanford Encyclopedia of Philosophy*, 1999.

Hindle, Brooke. *The Pursuit of Science in Revolutionary America, 1735-1789*. New York: W. W. Norton, 1974.

Hopkins, Terence K. and Immanuel Wallerstein. "Patterns of Development of the Modern World-System." *Review* 1 (2) (Fall 1977), 111-145.

Hughes, J. Donald. *An Environmental History of the World: Humankind's Changing Role in the Community of Life*. New York: Routledge, 2001.

———. "Global Dimensions of Environmental History." *Pacific Historical Review* 70 (1) (2001): 91-101.

———, ed. *The Face of the Earth: Environment and World History*. Armonk, N.Y.: M.E. Sharpe, 2000.

Huntington, Samuel P. *The Clash of Civilizations and the Remaking of World Order*. New Delhi: Viking, Penguin Books India (P) Ltd., 1996.

Husserl, Edmund. *Ideas Pertaining to a Pure Phenomenology and to a Phenomenological Philosophy* First Book. Translated by F. Kersten. The Hague: Martinus Nijhoff Publishers, 1983.

———. *Ideas Pertaining to a Pure Phenomenology and to a Phenomenological Philosophy* Second Book. Translated by R. Rojcewicz and A. Schuwer. Dordrecht: Kluwer Academic Publishers, 1989.

———. *Logical Investigation: Volume II*. Translated by J. N. Findlay. London: Routledge & Kegan Paul, 1970.

———. *On the Phenomenology of the Consciousness of Internal Time (1893-1917)*. Translated by John Brough. Dordrecht: Kluwer Academic Publishers, 1991.

Jackson, Kenneth T. *Crabgrass Frontier: The Suburbanization of the United States*. Oxford: Oxford University Press, 1985.

Jaffe, David. "The Village Enlightenment in New England, 1760-1820." *William and Mary Quarterly, 3rd Series*. 47 (1990): 327-346.

John, Richard. "Recasting the Information Infrastructure for the Industrial Age." In Alfred D. Chandler Jr. and James W. Cortada, *A Nation Transformed by Information: How Information has Shaped the United States from Colonial Times to the Present*. New York: Oxford University Press, 2000.

John, Richard R. *Spreading the News: The American Postal System from Franklin to Morse*. Cambridge, Mass.: Harvard University Press, 1995.

Jones, Eric L. *The European Miracle: Environments, Economies, and Geopolitics in the History of Europe and Asia*. 2nd edition. Cambridge: Cambridge University Press, 1987.

Juma, Calestous. *The Gene Hunters: Biotechnology and the Scramble for Seeds*. Princeton, N.J.: Princeton University Press, 1989.

Kaldor, Mary. *The Imaginary War: Understanding the East-West Conflict*. Oxford and Cambridge, Mass.: Blackwell, 1990.

Kant, Immanuel. *Critique of Pure Reason*. Translated by Norman Kemp Smith. New York: St. Martin's Press, 1965.

Keiderling, Gerhard. *"Rosinenbomber" über Berlin: Währungsreform, Blockade, Luftbrücke, Teilung*. Berlin: Dietz Verlag, 1998.

Keller, Evelyn Fox. A Feeling for the Organism: the Life and Work of Barbara McClintock. San Francisco: W. H. Freeman, 1983.

Kerényi, Carl, *Dionysos: Archetypal Image of Indestructible Life*. Translated by Ralph Manheim, Princeton, N.J.: Princeton University Press, 1976.

King, Roger J. H. "Environmental Ethics and the Built Environment." *Environmental Ethics* 22 (2000): 115-131.

Koertge, Noretta, ed. *A House Built on Sand*. Oxford: Oxford University Press, 1998.

Koop, Volker. *Kein Kampf um Berlin? Deutsche Politik zur Zeit der Berlin-Blockade 1948/1949*. Bonn: Bouvier Verlag, 1998.

———. "Die perfekte Blockade? Über Mythos und Legendenbildung." In *Auftrag Luftbrücke: Der Himmel über Berlin 1948-1949*, edited by Uwe Förster, Stephanie V. Hochberg, Ulrich Kubisch, and Dietrich Kuhlgatz, Berlin: Nicolai, 1998, 361-369.

Kopstein, Jeffrey. *The Politics of Economic Decline in East Germany, 1945-1989*. Chapel Hill and London: University of North Carolina Press, 1997.

Koschnik, Albrecht. "The Democratic Societies of Philadelphia and the Limits of the American Public Sphere, Circa 1793-1795," *William and Mary Quarterly* 3rd Series 58 (July, 2001): 615-636.

Kuhn, Thomas S. *The Structures of Scientific Revolutions*. Chicago: University of Chicago Press, 1970.

LaFreniere, Gilbert F. "Rousseau and the European Roots of Environmentalism," *Environmental History Review* 14:4 (Winter 1990): 41-72.

Landes, David. *The Wealth and Poverty of Nations: Why Some Are So Rich and Some So Poor*. New York: W. W. Norton, 1998.

Lemisch, Jesse. "Jack Tar in the Streets: Merchant Seamen in the Politics of Revolutionary America," in *In Search of Early America: The William and Mary Quarterly 1943-1993*. Richmond Virginia: William Byrd Press, 1993.

Longino, Helen. *Science as Social Knowledge: Values and Objectivity in Scientific Inquiry*. Princeton, N.J.: Princeton University Press, 1990.

———. *The Fate of Knowledge*. Princeton, N.J.: Princeton University Press, 2002.

Mark, David M., Barry Smith, and Barbara Tversky. "Ontology and Geographic Objects: An Empirical Study of Cognitive Categorization." Heidelberg: Springer Verlag (1999). http://link.springer.de/link/service/series/0558/bibs/1661/16610283.htm.

Marx, Karl. *Capital*. Vol. I. New York: Vintage, 1977.

———. *The Grundrisse*. Translated by M. Nicolaus. New York: Vintage, 1973.

Marx, Leo. *The Machine in the Garden: Technology and the Pastoral Ideal in America*. Oxford: Oxford University Press, 1964.

Masefield, G.B. "Crops and Livestock." In *The Cambridge Economic History of Europe IV: The Economy of Expanding Europe in the Sixteenth and Seventeenth Centuries*, edited by E. E. Rich and C. H. Wilson. London: Cambridge University Press, 1967, 276-301.

Mayumi, Kozo. "Temporary Emancipation from the Land: From the Industrial Revolution to the Present." *Ecological Economics* 4 (1) (1991): 35-56.

McEwen, Indra Kagis. *Socrates' Ancestor: An Essay on Architectural Beginnings*. Cambridge, Mass.: MIT Press, 1993.

McLaughlin, Mary-Beth, "Rural relaxation, urban amenities: Many developers consider Swanton northwest Ohio's next hot spot." *The Blade* (Toledo, Ohio), April 29, 2001, H1.

McNeill, J. R. *Something New Under the Sun: An Environmental History of the Twentieth Century*. New York: W. W. Norton, 2000.

Merleau-Ponty, Maurice. *Phenomenology of Perception*. Translated by Colin Smith. London: Routledge & Kegan Paul, 1962.

———. *The Visible and the Invisible: Followed by Working Notes*. Translated by Alphonso Lingis. Evanston, Ill.: Northwestern University Press, 1968.

Minkowski, Philip and Esther-Mirjam Sent, eds. *Science Bought and Sold: Essays in the Economics of Science*. Chicago: University of Chicago Press, 2002.

Mintz, Sidney W. *Sweetness and Power*. New York: Penguin, 1985.

Miskimin, Harry. *The Economy of Early Renaissance Europe*. Cambridge: Cambridge University Press, 1975.

Mitcham, Carl. *Thinking through Technology. The Path between Engineering and Philosophy*. Chicago: The University of Chicago Press, 1994.

Moore, Jason W. "Environmental Crises and the Metabolic Rift in World-Historical Perspective." *Organization & Environment* 13(2) (June 2000): 123-58.

———. "Sugar and the Expansion of the Early Modern World-Economy." In *Review: A Journal of the Fernand Braudel Center* 23 (3) (2000): 409-433.

———. "*The Modern World-System* as Environmental History? Transformations of Capital and Nature over the *Longue Duree*." Paper presented to the 25th annual conference of the Political Economy of the World-System section, American Sociological Association, Blacksburg, Va., April 17-19, 2001.

———. "The Crisis of Feudalism: An Environmental History." *Organization & Environment* 15(3) (September 2002): 296-317.

Mott, Frank Luther. *American Journalism: A History of Newspapers in the United States Through 250 Years, 1690-1940*. New York: The Macmillan Company, 1941.

Murphy, David E., and Sergei A. Kondrashev. *Battleground Berlin: CIA vs. KGB in the Cold War*. New Haven, Conn.: Yale University Press, 1999.

Nicolescu, Basarab. *Manifesto of Transdisciplinarity*. Translated by Karen-Claire Voss. Albany: State University of New York Press, 2002.

Oakes, Guy. *The Imaginary War: Civil Defense and American Cold War Culture*. New York and Oxford: Oxford University Press, 1994.

Owen, G. E. L. "Aristotle on Time." In *Articles on Aristotle*, Vol. 3 "Metaphysics," edited by J. Barnes. London: Duckworth, 1979.

Paisley, Jeffrey L. "*The Tyranny of Printers*": *Newspaper Politics in the Early American Republic*. Charlottesville: University Press of Virginia, 2001.

Pepper, Steven C. *World Hypotheses*. Berkeley: University of California Press, 1942.

Perlin, Frank. *A Forest Journey: The Role of Wood in the Development of Civilization*. Cambridge, Mass.: Harvard University Press, 1989.

Peskin, Lawrence A. "From Protection to Encouragement: Manufacturing and Mercantilism in New York City's Public Sphere, 1783-1795." *Journal of the Early Republic* 18 (Winter, 1998): 589-615.

Petroski, Henry. *The Evolution of Useful Things*. New York: Vantage Books, 1994.

Plutarch, *Lives, Vol. I*. Translated by Bernadotte Perrin. Cambridge, Mass.: Harvard University Press, 1914, reprinted 1998.

Polyani, Michael. *Personal Knowledge*. New York: Harper and Row Publishers, 1964.

Ponting, Clive. *A Green History of the World*. New York: St. Martin's Press, 1991.

Postan, M. M. *Medieval Economy and Society*. New York: Pelican, 1972.

Powers, Thomas M., and Paul Kamolnick, eds. *From Kant to Weber: Freedom and Culture in Classical German Social Theory*. Malabar, Fla.: Krieger Publishing Company, 1999.

Prince, Carl E. et al., eds. *The Papers of William Livingston*. New Brunswick, N.J.: Rutgers University Press, 1979.

Raina, Dhruv. "Beyond the Diffusionist History of Colonial Science." *Social Epistemology* 12, no. 2 (1998): 203-213.

———. "Reconfiguring the Centre: The Structure of Scientific Exchanges Between Colonial India and Europe." *Minerva* 34 (1996): 9-42.

———. "The Present in the Past: Trajectories for the Social History of Science." *India: Another Millenium?*, edited by Romila Thapar. New Delhi: Viking, 2000.

———. "The Young P.C. Ray and the Inauguration of the Social History of Science in India." *Science, Technology and Society* 2, no. 1 (1997): 1-40.

Raina, Dhruv and Irfan Habib. "The Unfolding of an Engagement: The Dawn on Science, Technical Education and Industrialisation." *Studies in History* 9 no. 1 (1995): 95-117.

———. "The Moral Legitimation of Science: Bhadrolok Reflections on the Theory of Evolution." *Social Studies of Science* 26 (1996): 9-42.

Rediker, Marcus. *Between the Devil and the Deep Blue Sea.* New York: Cambridge University Press, 1987.

Remer, Rosalind. *Printers and Men of Capital: Philadelphia Book Publishers in the New Republic.* Philadelphia: University of Pennsylvania Press, 1996.

Richards, John F., and Richard P. Tucker, eds. *World Deforestation in the Twentieth Century.* Durham, N.C.: Duke University Press, 1988.

Rome, Adam. *The Bulldozer in the Countryside: Suburban Sprawl and the Rise of American Environmentalism.* Cambridge: Cambridge University Press, 2001.

Romm, James. *Herodotus.* New Haven, Conn.: Yale University Press, 1998.

Rousseau, Jean-Jacques. *Confessions.* Translated by J. M. Cohen. New York: Penguin, 1953.

———. *Reveries of the Solitary Walker.* Translated by Peter France. New York: Penguin, 1979.

———. *The First and Second Discourses.* Edited and Translated by Victor Gourevitch. New York: Harper and Row, 1986.

Sauer, Carl. "Destructive Exploitation in Modern Colonial Expansion." *Comptes Rendus du Congres International de Geographie* (Amsterdam), vol. 2, sec. 3c (1938): 494-499.

———. *Selected Essays, 1963-1975.* Berkeley, Calif.: Turtle Island Foundation, 1981.

———. "Theme of Plant and Animal Destruction in Economic History." *Journal of Farm Economics* 20 (8) (1938): 765-775.

Schlegelmilch, Arthur. *Hauptstadt im Zonendeutschland: Die Entstehung der Berliner Nachkriegsdemokratie 1945-1949.* Berlin: Haude & Spener, 1993.

Schliermacher, Friedrich. *On Religion, Speeches to its Cultured Despisers.* Translated by Richard Crouter. Cambridge: Cambridge Univesity Press, 1988.

Schlosberg, David. *Environmental Justice and the New Pluralism.* Oxford: Oxford University Press, 1999.

Schneider, Marcel. *Jean-Jacques Rousseau et l'espoir écologiste.* Paris: Éditions Pygmalion, 1978.

Schutz, Alfred and Thomas Luckmann. *The Structures of the Life-World.* Translated by Richard M. Zaner and H. Tristram Engelhardt, Jr. Evanston, Ill.: Northwestern University Press, 1973.

Seamon, David, ed. *Dwelling, Seeing and Designing: Toward a Phenomenological Ecology.* Albany: State University of New York Press, 1993.

Sellers, Christopher. "Body, Place and the State: The Makings of an 'Environmentalist' Imaginary in the Post-World War II U.S." *Radical History Review* 74 (1999): 31-64.

Scott, James C. *Seeing Like A State: How Certain Schemes to Improve the Human Condition Have Failed.* New Haven, Conn.: Yale University Press, 1998.

Shields, David. *Civil Tongues and Polite Letters in British America.* Chapel Hill: University of North Carolina Press, 1997.

Shrader-Frechette, K. *Ethics of Scientific Research.* Lanham, Md.: Rowman and Littlefield, 1994.

Simmel, Georg. *On Individuality and Social Formal: Selected Writings.* Edited and with an introduction by Donald N. Levine. Chicago: The University of Chicago Press, 1971.

Simons, Peter. *Parts: a Study in Ontology.* Oxford: Clarendon Press, 1987.

Sinclair, Bruce. *Philadelphia's Philosopher Mechanics: A History of the Franklin Institute, 1824-1865.* Baltimore: Johns Hopkins University Press, 1974.

Smil, Vaclav. *Energies: An Illustrated Guide to the Biosphere and Civilization*. Cambridge, Mass.: MIT Press, 1999.

Smith, Barry. "Fiat Objects." *Topoi* 20 (2001): 131-148.

Smith, David A. "Uneven Development and the Environment: Toward a World-System Perspective." *Humboldt Journal of Social Relations* 20 (1) (1994): 151-75.

Smith, Neil. *Uneven Development*. Oxford: Blackwell, 1984.

Smith, Paul H. ed., *Letters of Delegates to Congress*. Washington, D.C.: Library of Congress, 1976–2003.

Soja, Edward W. *Postmodern Geographies*. London: Verso, 1989.

Southern Poverty Law Center. "The Civil Rights Memorial." Montgomery, Alabama.

Steege, Paul. "More Than an Airlift: Constructing the Berlin Blockade as a Cold War Battle, 1946-1949." Ph.D. diss., University of Chicago, 1999.

Stephanson, Anders. "United States and the Cold War." In *The Origins of the Cold War in Europe: International Perspectives*, edited by David Reynolds. New Haven, Conn., and London: Yale University Press, 1994, 23-52.

Stern, Robert A. M. *Pride of Place: Building the American Dream*. Boston and New York: Houghton Mifflin and American Heritage, 1986.

Stern, Steve J. "Feudalism, Capitalism, and the World-System in the Perspective of Latin America and the Caribbean." *American Historical Review* 93 (Oct. 1988): 829-873.

Stivers, William. "The Incomplete Blockade: Soviet Zone Supply of West Berlin, 1948-49." *Diplomatic History* 21, no. 4 (Fall 1997): 569-602.

Szcygielski, Wocjiech. "Die Okonomische Aktivitat de Polnischen Adels im 16.-18. Jarhumdert." *Studia Historia Economicae* 2 (1967): 83-101.

Tagore, Rabindranath. *The Religion of Man*. New Delhi: Rupa & Co, 2002 Hard Cover edition.

The Encylopedia Brittanica, Eleventh Edition, "Herodotus." Cambridge: Cambridge University Press, 1910-1911.

Thomas, William L., Jr., edited by *Man's Role in Changing the Face of the Earth*. Chicago: University of Chicago Press, 1956.

Thomasson, Amie L. "Geographic Objects and the Science of Geography," *Topoi* 20 (2001): 149-159.

Tilly, Charles. "Food Supply in Early Modern Europe." In *The Formation of National States in Early Modern Europe*, edited by Charles Tilly. Princeton, N.J.: Princeton University Press, 1975, 380-455.

Timmons, Robert J., and Peter Grimes. "Extending the World-System to the Whole System: Toward a Political Economy of the Biosphere." In *Ecology and the World-System*, edited by Walter L. Goldfrank, Andrew Szasz, and David Goodman. Westport, Conn.: Greenwood Press, 1999, 59-83.

Tucker, Richard P. and John F. Richards, eds. *Global Deforestation and the Nineteenth Century World Economy*. Durham, N.C.: Duke University Press, 1983.

Turner, Victor. *Blazing the Trail: Way Marks in the Exploration of Symbols*. Edited by Edith Turner. Tucson and London: University of Arizona Press, 1992.

———. *The Forest of Symbols: Aspects of Ndembu Ritual*. Ithaca, N.Y., and London: Cornell University Press, 1973.

Turner II, B.L., William C. Clark, Robert W. Kates, John F. Richards, Jessica T. Mathews, and William B. Meyer, eds. *The Earth as Transformed by Human Action: Global and Regional Changes in the Biosphere over the Past 300 Years*. Cambridge: Cambridge University Press, 1990.

Tusa, Ann and John Tusa. *The Berlin Airlift*. New York: Athenaeum, 1988.

Utterstrom, Gustaf. "Climate Fluctuations and Population Problems in Early Modern Europe." *Scandinavian Economic History Review* 3 (1) (1955): 3-47.

Varzi, Achile C. "Vagueness in Geography." *Topoi* 20: 119-30.

———. "Words and Objects." *Individuals, Essence, and Identity. Themes of Analytic Metaphysics*. Edited by Andrea Bottani, Massimiliano Carrara, and Daniele Giaretta. Dordrecht: Kluwer Academic Publishers, 2002.

Veblen, Thorstein. *The Theory of the Leisure Class*. New York: Mentor, 1953.

Waldstreicher, David. *In the Midst of Perpetual Fetes: The Making of American Nationalism, 1776-1829*. Chapel Hill: University of North Carolina Press, 1997.

Wallerstein, Immanuel. *The Modern World-System I: Capitalist Agriculture and the Origins of the European World-Economy in the Sixteenth Century*. New York: Academic Press, 1974.

———. *The Modern World-System II*. New York: Academic Press, 1980.

———. *The Modern World-System III*. New York: Academic Press, 1989.

Warner, Michael. *The Letters of the Republic: Publication and the Public Sphere in Eighteenth-Century America*. Cambridge, Mass.: Harvard University Press, 1990.

Webb, Walter Prescott. *The Great Frontier*. Austin: University of Texas Press, 1964.

Westoby, Jack. *Introduction to World Forestry*. Oxford: Basil Blackwell, 1989.

Wheatley, Helen, ed. *Agriculture, Resources Exploitation, and Environmental Change*. Aldershot, UK, and Brookfield, Vt.: Variorum/Ashgate Publishing Company, 1997.

White, Leonard. *The Federalists: A Study in Administrative History, 1789-1801*. New York: The Free Press, 1948.

Wiggins, David. *Sameness and Substance*. Oxford: Blackwell, 1980.

Wolf II, Edwin. *The Book Culture of a Colonial American City: Philadelphia Books, Bookmen, and Booksellers*. New York: Oxford University Press, 1988.

Worster, Donald, ed. *The Ends of the Earth: Perspectives on Modern Environmental History*. Cambridge: Cambridge University Press, 1988.

———. "Transformations of the Earth: Toward an Agroecological Perspective in History," *Journal of American History* 76 (4) (March 2001): 1087-1106.

———. "World without Borders: The Internationalizing of Environmental History." In *Environmental History*, edited by Kendall E. Bailes. Lanham, Md.: University Press of America, 1985, 661-669.

Wood, D. and R. Beck. "Tour Personality: The Independence of Environmental Orientation in Interpersonal Behavior." *Journal of Environmental Psychology* 10 (3) (September 1990): 177-207.

Wright, John Kirtland. *Human Nature in Geography: Fourteen Papers, 1925-1965*. Cambridge, Mass.: Harvard University Press, 1966.

Index

About the Contributors

Gary Backhaus received his Ph.D. in philosophy in 1992 from the American University. He currently teaches at Morgan State University in Baltimore, Maryland. His research concerns interdisciplinary, transdisciplinary, and applied phenomenology. Recent book publications include *Lived Images: Mediations in Experience, Life-World and I-hood*, co-edited with Matti Itkonen, and *Tensional Landscapes: The Dynamics of Boundaries and Placements*, co-edited with John Murungi. Recent articles appear in *Human Studies*, *Analecta Husserliana*, and *Culture & Society: International Journal of the Human Sciences*.

Deepwanita Dasgupta received her Ph.D. in philosophy in 1995 from the Jadavpur University in Calcutta. She is currently pursuing a second doctorate in the philosophy of science at the University of Minnesota. Her main interests include the philosophy of science, social science, cognitive science, and the historiography of science.

Robert Kirkman received his Ph.D. in philosophy from the State University of New York at Stony Brook. He is Assistant Professor in the School of Public Policy of the Georgia Institute of Technology. His research concerns environmental philosophy with a special focus on the ethical issues raised by metropolitan growth in the United States. He is author of *Skeptical Environmentalism: The Limits of Philosophy and Science*. His recent articles have appeared in *Environmental Ethics* and *Journal of Value Inquiry*.

Jason Moore is a world historian and graduate student in geography at the University of California, Berkeley. His articles and short essays concerning globalization, ecology, and labor have appeared in *Antipode: A Journal of Radical Geography*, *Capitalism Nature Socialism*, *Critical Sociology*, *Journal of World-Systems Research*, *Monthly Review*, *Organization & Environment*, *Review: A Journal of the Ferdnand Braudel Center*, *Science & Society*, *Theory and Society*, and (with Giovanni Arrighi) *Phases of Capitalist Development*. In 2002 he was given a Distinguished Scholarship Award by the Political Economy of the World-System section, American Sociological Association, for his articles on world environmental history. At present he is writing an environmental history of the transition from feudalism to capitalism.

François-Xavier Nzi iyo nsenga received his Maîtrise ès sciences (M. sc.) (Sociologie) in 2001 from the Université de Monteal and, from the same institution, received his Maîtrise ès sciences appliquées (M. sc. A.) (Aménagement) in 1979. He is an independent scholar in design and sociology. His main research interest involves prospective sociocultural ramifications in product design.

Lawrence Peskin received his Ph.D. in history from the University of Maryland in 1998. He is Assistant Professor of History at Morgan State University. His current research concerns the history of America's late eighteenth-century relations with Algiers. He is author of *Manufacturing Revolution: The Intellectual Origins of Early American Industry*.

John M. Rose received his Ph.D. from Georgetown University. He is Professor of Philosophy at Goucher College in Maryland. He is author of many scholarly papers on contemporary continental philosophy, the Presocratics, Neoplatonism, and the application of philosophy to lived experiences. His recent research takes up the question of the concept of time in Ancient Greece and German philosophy. His most recent publications are "Multiple Truths and Multiple Narratives: Nietzsche's Perspectivism and the Narrative Structure of *The Alexandria Quartet*" in *Lawrence Durrell and the Greek World*, and "Athens, Alexandria, Jerusalem: Heidegger and Levinas on The Other and the City" in *Tensional Landscapes: The Dynamics of Boundaries and Placements*. Professor Rose is the convener for both the Society for Realist/Antirealist Discussion and the yearly meetings of the Society for the Contemporary Assessment of Platonism with the American Philosophical Association.

Dennis E. Skocz received his Ph.D. in philosophy at Duquesne University in 1974. He is currently an independent scholar and is a U.S. diplomat currently assigned to the Pentagon to work with NATO. His articles have appeared in *Journal of the International Association of Environmental Philosophy*, *Phenomenological Inquiry*, and *Proceedings of the North American Heidegger Conference*. He also has an article "Earthling or Cosmopolitan? The Limits of Interlocution" in *Tensional Landscapes: The Dynamics of Boundaries and Placements*, which is edited by Backhaus and Murungi.

Paul Steege received his Ph.D. in modern European history from the University of Chicago in 1999. He is currently Assistant Professor of History at Villanova University. His research focuses on twentieth-century Germany, the Cold War, and the history of everyday life. His book *Between War and Peace: The Battle for Berlin, 1946-1949* is forthcoming.

www.ingramcontent.com/pod-product-compliance
Ingram Content Group UK Ltd.
Pitfield, Milton Keynes, MK11 3LW, UK
UKHW012039150926
475560UK00001B/84

9 780739 107645